Modeling and Simulation of Fluid Flow and Heat Transfer

In the rapidly advancing modern world, scientific and technological understanding and innovation are reaching new heights. Computational fluid dynamics and heat transfer have emerged as powerful tools, playing a pivotal role in the analysis and design of complex engineering problems and processes. With the ability to mathematically model various engineering phenomena, these computational tools offer a deeper understanding of intricate dynamics before the physical prototype is created. Widely employed as simulation tools, computational fluid dynamics and heat transfer codes enable the virtual or digital prototype development of products and devices involving complex transport and multiphasic phenomena. They have become an indispensable element of the agile product development environment across diverse sectors of manufacturing, facilitating accelerated product development cycles.

Key features of this book:

- Covers the analysis of advanced thermal engineering systems.
- Explores the simulation of various fluids with slip effect.
- Applies entropy and optimization techniques to thermal engineering systems.
- Discusses heat and mass transfer phenomena.
- Explores fluid flow and heat transfer in porous media.
- Captures recent developments in analytical and computational methods used to investigate the complex mathematical models of fluid dynamics.
- Covers the application of mathematical and computational modeling techniques to fluid flow problems in various geometries.

Modeling and Simulation of Fluid Flow and Heat Transfer delves into the fascinating world of fluid dynamics and heat transfer modeling, presenting an extensive exploration of these subjects. This book is a valuable resource for researchers, engineers, and students seeking to comprehend and apply numerical methods and computational tools in fluid dynamics and heat transfer problems.

Engineering Tribology, Manufacturing and Applied Energy

Series Editors: Yashvir Singh and Nishant Kumar Singh

About The Series:

Tribology is the study of parts when they are in contact with each other and have a considerable impact during their sliding or rotating motion. This series will publish the books highlighting the latest developments and research in tribology with its applications in the manufacturing and energy sectors. This series will include books covering established areas of tribology and emerging fields such as tribo-chemistry, surface engineering, nano-tribology, and bio-tribology. This series will focus on the recent developments and advancements in green tribology, for example, environmentally friendly composites. Aimed at senior undergraduate and graduate students, academic researchers, and professionals, this series will cover the latest techniques in the field of tribology with applications in diverse fields such as manufacturing, energy, and medicine.

Modeling and Simulation of Fluid Flow and Heat Transfer
Edited by Reshu Gupta and Mukesh Kumar Awasthi

For more information on series page, please visit our website: www.routledge.com/Engineering-Tribology-Manufacturing-and-Applied-Energy/book-series/CRCETMAE

Modeling and Simulation of Fluid Flow and Heat Transfer

Edited by
Reshu Gupta
and Mukesh Kumar Awasthi

CRC Press
Taylor & Francis Group
Boca Raton London New York

CRC Press is an imprint of the
Taylor & Francis Group, an **informa** business

First edition published 2024
by CRC Press
6000 Broken Sound Parkway NW, Suite 300, Boca Raton, FL 33487-2742

and by CRC Press
4 Park Square, Milton Park, Abingdon, Oxon, OX14 4RN

CRC Press is an imprint of Taylor & Francis Group, LLC

ISBN: 978-1-032-70606-1 (hbk)
ISBN: 978-1-032-71206-2 (pbk)
ISBN: 978-1-032-71207-9 (ebk)

DOI: 10.1201/9781032712079

Typeset in Sabon
by codeMantra

Contents

Preface

The study of fluid flow and heat transfer has emerged as a critical area of investigation across a diverse range of technical and scientific disciplines. From designing efficient energy systems to advancing cutting-edge technologies, the accurate representation and simulation of these intricate phenomena hold paramount importance. In light of this significance, we present the book *Modeling and Simulation of Fluid Flow and Heat Transfer* as an extensive resource catering to individuals, both students and professionals, with an avid interest in this captivating field.

Fluid flow and heat transfer encompass complex processes governed by interconnected physical principles and mathematical relationships. Understanding and predicting fluid and thermal behavior in various systems assume crucial significance in the pursuit of optimizing performance, ensuring safety, and fostering innovation. With the advent of powerful computational tools and sophisticated numerical techniques, modeling and simulation have become indispensable instruments for studying and analyzing these phenomena.

This book adopts a well-balanced approach to the modeling and simulation of fluid flow and heat transfer, aiming to bridge the gap between theoretical concepts and practical applications. It comprehensively addresses the fundamental principles of fluid mechanics, thermodynamics, and heat transfer, providing readers with a robust foundation to tackle real-world engineering challenges. This book embarks on a systematic exploration of mathematical modeling and numerical techniques, allowing readers to develop a methodical understanding of the subject. It commences by elucidating the fundamental equations governing fluid flow and heat transfer, such as the Navier–Stokes equations, energy conservation equations, and pertinent boundary conditions. Subsequently, it delves into the process of converting physical systems into mathematical models, highlighting the significance of assumptions, simplifications, and empirical correlations required to facilitate accurate simulations. The resulting transformed equations are solved utilizing diverse numerical methods, semi-analytical techniques, and computational fluid dynamics (CFD), with the outcomes presented in graphical form and concisely explicated.

This book consists of 13 chapters, each delving into distinct aspects of fluid flow and heat transfer. Chapter 1 focuses on the analysis of nanofluid flow and heat transfer between two parallel plates, employing the homotopy perturbation and Akbari Ganji methods. Chapter 2 is dedicated to examining coal gasification, involving the combustion of coal and the extraction of produced syngas through drilled wells. Chapter 3 investigates the thermal and flow behaviors of MXene (Ti3C2)–dispersed blood flowing through curved arteries with cosine-shaped stenosis. Subsequent chapters delve into topics such as tube and shell heat exchangers, boundary layer flow problems, magnetic fluid flow, magnetohydrodynamic flows, blast wave propagation, isentropic fluid flow, Rayleigh–Taylor instability, non-Newtonian flow, and nanofluid flow over porous surfaces, among others.

This book primarily targets undergraduate and graduate students pursuing degrees in various engineering disciplines. In addition, it caters to researchers actively engaged in the study of fluid dynamics and heat transfer, as well as professionals seeking a comprehensive reference guide within this domain. The overarching objective of *Modeling and Simulation of Fluid Flow and Heat Transfer* is to provide a valuable resource that enables readers to effectively comprehend the intricacies associated with fluid flow and heat transfer phenomena. It is anticipated that this book will facilitate the exploration of novel perspectives, thereby fostering avenues for further discovery and innovation.

In conclusion, we hope that this comprehensive and meticulously designed book will contribute significantly to the understanding and advancement of fluid flow and heat transfer studies. We extend our gratitude to all the contributors and reviewers who have been instrumental in shaping this endeavor. We are optimistic that this book will inspire readers to unravel the complexities of fluid dynamics and heat transfer, leading to innovative solutions and transformative breakthroughs in diverse fields of engineering and scientific research.

Dr. Reshu Gupta

Dr. Mukesh Kumar Awasthi

Editors

Dr. Reshu Gupta did her Ph.D. on the topic "Study of some problems of heat transfer in the flows of some fluids." She is an Assistant Professor in the Applied Science Cluster (Mathematics) at the University of Petroleum and Energy Studies (UPES) in Dehradun, India. She has 20 years teaching experience. She has taught Linear Algebra, Fluid Dynamics, Numerical Methods, Differential Equations, Calculus, Abstract Algebra, Ring Theory, and Discrete Mathematics. Her research areas include Fluid Dynamics, Differential Equations, and Heat and Mass Transfer. She has published several papers in various journals and conference proceedings.

Dr. Mukesh Kumar Awasthi did his Ph.D. on the topic "Viscous correction for the potential flow analysis of capillary and Kelvin-Helmholtz instability." He is an Assistant Professor in the Department of Mathematics at Babasaheb Bhimrao Ambedkar University, Lucknow. Dr. Awasthi has specialized in the mathematical modeling of flow problems. He has taught courses in Fluid Mechanics, Discrete Mathematics, Partial Differential Equations, Abstract Algebra, Mathematical Methods, and Measure Theory to postgraduate students. He is well versed in the mathematical modeling of flow problems, and he can solve these problems analytically as well as numerically. He has a good grasp of the subjects such as viscous potential flow, electro-hydrodynamics, magneto-hydrodynamics, heat, and mass transfer.

Dr. Awasthi has qualified for the National Eligibility Test (NET) conducted at All-India level in 2008 by the Council of Scientific and Industrial Research (CSIR) and received Junior Research Fellowship (JRF) and Senior Research Fellowship (SRF) for doing his research. He has published 115 plus research publications (journal articles/books/book chapters/conference articles) with Elsevier, Taylor & Francis, Springer, Emerald, World Scientific, and many other national and international journals and conferences. Moreover, he has published seven books. He has attended many symposia, workshops, and conferences in mathematics as well as fluid mechanics. He has received the "Research Awards" four consecutive times, from 2013 to 2016 from the

University of Petroleum and Energy Studies, Dehradun, India. He has also received the start-up research fund for his project "Nonlinear study of the interface in multilayer fluid system" from UGC, New Delhi. He is listed in the top 2% influential researchers in the World prepared by Stanford University based on Scopus data in 2022.

Contributors

Walid Abouloifa
Sultan Moulay Slimane University of Beni Mellal
Polydisciplinary Faculty of Khouribga, Research Team of Energy, Materials, Atomics and Information Fusion
Main Khouribga, Morocco

J. Ahuja
Department of Mathematics
PGGC
Chandigarh, India

Anupam Bhandari
Department of Mathematics
University of Petroleum and Energy Studies
Dehradun, India

Ankita Bisht
Department of Mathematics & Scientific Computing, School of Physical Sciences
Amity University
Punjab, India

Arvind Singh Bisht
Centre for Energy Studies
NIT
Hamirpur, India

Divya Chaturvedi
Department of Engineering Mathematics and Computing
Madhav Institute of Technology and Science
Gwalior, India

Amine El Harfouf
Sultan Moulay Slimane University of Beni Mellal
Polydisciplinary Faculty of Khouribga, Research Team of Energy, Materials, Atomics and Information Fusion
Main Khouribga, Morocco

P. Girotra
Department of Mathematics
RGGC
Ambala, India

Nidhi Handa
Department of Mathematics
Kanya Gurukula Mahavidyalaya
Haridwar, India

Sanaa Hayani Mounir
Sultan Moulay Slimane University of Beni Mellal
Polydisciplinary Faculty of Khouribga, Research Team of Energy, Materials, Atomics and Information Fusion
Main Khouribga, Morocco

Rachid Herbazi
ENSAT, Abdelmalek Esaàdi University
Tangier, Morocco
and
ERCMN, FSTT, Abdelmalek Esaàdi University
Tangier, Morocco
and
LSIA, EMSI
Tangier, Morocco

Manjunathayya Holliyavar
Department of Mathematics
K.L.E. Society's S.S.M.S College
Athani, Karnataka, India

Akmal Husain
Department of Mathematics
University of Petroleum and Energy Studies
Dehradun, India

Ekta Jain
Department of Mathematics
University of Delhi
Delhi, India

Najwa Jbira
Sultan Moulay Slimane University of Beni Mellal
Polydisciplinary Faculty of Khouribga
Research Team of Energy, Materials, Atomics and Information Fusion
Main Khouribga, Morocco

Amit Kumar
Department of Mathematics
University of Petroleum & Energy Studies
Dehradun, India

J. K. Madhukesh
Department of Mathematics
Amrita School of Engineering
Amrita Vishwa Vidyapeetham
Bengaluru, India

Sanjalee Maheshwari
Department of Mathematics & Scientific Computing
NIT
Hamirpur, India

Hassane Mes-Adi
Laboratoire d'ingénierie des Procédés Informatiques et Mathématiques
École nationale des sciences Appliquées Khouribga
Université Sultan Moulay Slimane Béni Mellal
Khouribga, Morocco

Bhagyashri Patgiri
Department of Mathematics
Cotton University
Guwahati, Assam, India

Ashish Paul
Department of Mathematics
Cotton University
Guwahati, India

Minakshi Poonia
Department of Engineering Mathematics and Computing
Madhav Institute of Technology and Science
Gwalior, India

Shailandra Kumar Prasad
Department of Mechanical Engineering
National Institute of Technology
Jamshedpur, India

S. D. Ram
Department of Applied Mathematics and Humanities
SBP Government Polytechnic Azamgarh
Azamgarh, India

G. K. Ramesh
Department of Mathematics
K.L.E. Society's J.T. College
Gadag, Karnataka, India

Niraj Rathore
Department of Mathematics
Central University of Karnataka
Kalaburagi, India

Atul Kumar Ray
Department of Engineering Mathematics and Computing
Madhav Institute of Technology and Science
Gwalior, India

Yassine Roboa
Sultan Moulay Slimane University of Beni Mellal
Polydisciplinary Faculty of Khouribga
Research Team of Energy, Materials, Atomics and Information Fusion
Main Khouribga, Morocco

N. Sandeep
Department of Mathematics
Central University of Karnataka
Kalaburagi, India

Ankur Kumar Sarma
Department of Mathematics
Cotton University
Guwahati, India

Shakuntla Sharma
Department of Applied Sciences
Tula's Institute
Dehradun, India

Devansh Shrivastava
Department of Petroleum Engineering
University of Aberdeen
Scotland, United Kingdom

Dhanpal Singh
Department of Mathematics
Keshav Mahavidyalaya
University of Delhi
Delhi, India

Mithilesh Singh
Department of Mathematics
Prof. Rajendra Singh (Rajju Bhaiya) Institute of Physical Sciences for Study and Research, Veer Bahadur Singh
Purvanchal University
Jaunpur, India

Mrityunjay Kumar Sinha
Department of Mechanical Engineering
National Institute of Technology
Jamshedpur, India

Dig Vijay Tanwar
Department of Mathematics
Graphic Era (Deemed to be University)
Dehradun, India

B. Vasu
Department of Mathematics
Motilal Nehru National Institute of Technology
Allahabad, India

K. Vinutha
Department of Studies in Mathematics
Davangere University
Davangere, India

Abderrahim Wakif
Hassan II University
Faculty of Sciences Aïn Chock Laboratory of Mechanics
Mâarif, Casablanca, Morocco

Chapter 1

Flow behavior of a conducting fluid in an unsteady state with Brownian motion

Amine El Harfouf, Yassine Roboa, Sanaa Hayani Mounir
Sultan Moulay Slimane University of Beni Mellal

Hassane Mes-Adi
Université Sultan Moulay Slimane Béni Mellal

Walid Abouloifa and Najwa Jbira
Sultan Moulay Slimane University of Beni Mellal

Rachid Herbazi
Abdelmalek Esaàdi University

Abderrahim Wakif
Hassan II University

1.1 INTRODUCTION

The study of heat and mass transfer in the context of unstable compression of nanofluids between parallel plates has attracted significant attention in various industries and biological applications. This research area finds relevance in a wide range of fields, including chemical processing machinery, compression equipment, food processing systems, water cooling mechanisms, and polymer processing. Over the years, numerous researchers have contributed to advancing our understanding of this topic.

Stefan (1875) made pioneering contributions to the field by investigating the lubrication approximation for nanofluid flow and heat transfer between parallel plates. Building on this foundation, other studies explored related aspects, such as the magnetohydrodynamic compression flow of viscous fluid between infinite parallel plates (Siddiqui et al., 2008), suction or injection compression flow between parallel disks (Domairry & Aziz, 2009), and the compression flow of non-Newtonian fluids using second-grade fluids (Domairry & Aziz, 2009; Hayat et al., 2012). Additional research by Mahmood et al. (2007)

DOI: 10.1201/9781032712079-1

investigated compression flow across porous surfaces, where an increase in the Nusselt number with higher Prandtl numbers was observed, in agreement with findings by Mustafa et al. (2012). Sheikholeslami et al. (2014a) delved into the heat transfer and nanofluid flow between rotating parallel plates, noting a decrease in the Nusselt number with increasing magnetic and rotation parameters, as well as the Eckert number.

Nonlinearity is a common challenge encountered in heat transfer equations within the engineering domain. To address this complexity, researchers have employed various methodologies. These include the Differential Transformation Technique (DTM) (Jang et al., 2001; Sheikholeslami et al., 2015; Sheikholeslami & Ganji, 2015), Variation of Parameter Method (VPM) (Mohyud-Din & Yildirim, 2010), Homotopy Analysis Method (HAM) (Sheikholeslami et al., 2014b), Adomian Decomposition Method (ADM) (Mohyud-Din & Yildirim, 2010), and Homotopy Perturbation Method (HPM) (Sheikholeslami et al., 2014b). Each approach brings its own strengths and has been utilized in previous studies to tackle nonlinear heat transfer problems effectively.

The present research contributes to the current understanding by investigating the compression of nanofluids between parallel plates in the presence of a variable magnetic field. The study considers crucial factors, such as viscous dissipation and heat generation arising from shear-induced friction within the flow. To tackle the problem, a similarity transformation is employed to convert the governing equations into ordinary differential equations. These equations are then solved using the HPM, a powerful analytical technique introduced and enhanced by He (2004). The HPM offers distinct advantages, including convergence to accurate solutions through an infinite series representation, without the need for small parameters.

Furthermore, another promising semi-analytical technique, Akbari–Ganji's Method (AGM), has gained attention for its efficacy in studying nonlinear systems. Mirgolbabaee et al. (2016) successfully utilized AGM to investigate the dynamics of a Duffing-type nonlinear oscillator. In recent years, AGM has been increasingly employed by researchers to simulate nanofluid flow and heat transfer (Draper et al., 2022; El Harfouf et al., 2023; Sheikholeslami et al., 2017; Sheikholeslami & Ganji, 2016, 2017).

The primary objective of this study is to employ AGM to examine the phenomenon of unstable nanofluid flow between parallel plates. The investigation encompasses an in-depth analysis of key parameters, such as the squeezing number, magnetic number, volume fraction, Brownian motion coefficient, and thermophoretic diffusion, and their effects on the velocity and temperature profiles. By gaining insights into the behavior of nanofluid flow under compression, this research contributes to advancing the theoretical understanding and practical applications in various industries.

1.2 PROBLEM DESCRIPTION AND GOVERNING EQUATIONS

Figure 1.1 illustrates the configuration of the system consisting of two infinite parallel plates, where an incompressible viscous fluid undergoes unsteady two-dimensional squeezing. The distance between the plates, denoted as $h(t) = l(1-at)^{\frac{1}{2}}$, dynamically changes with time. For positive values of $\alpha(a>0)$, the plates gradually approach each other until they touch at $t = 1/\alpha$. Conversely, for negative values of $\alpha(0)$, the plates move apart. In addition, a variable magnetic field of strength $B(t) = B_0(1-at)^{\frac{-1}{2}}$ is applied along the y direction.

To account for the electromagnetic effects, the electric current and electromagnetic force are defined as $\vec{J} = \sigma\left(\vec{v}\times\vec{B}\right)$ and $\vec{F} = \vec{J}\times\vec{B}$, respectively. Consequently, it is determined that $\vec{F} = \sigma\left(\vec{v}\times\vec{B}\right)\times\vec{B}$.

This study takes into consideration both the viscous dissipation effect and heat generation resulting from shear-induced friction in the fluid flow. As a result, the governing equations describing the transfer of mass, momentum, energy, and species in the unsteady two-dimensional flow of a nanofluid are presented:

$$\vec{\nabla}\cdot\vec{V} = 0 \tag{1.1}$$

$$\rho\frac{\partial\vec{V}}{\partial t} + \rho\vec{V}\cdot\vec{\nabla}\vec{V} = -\vec{\nabla}p + \mu\nabla^2\vec{V} + \sigma\left(\vec{J}\times\vec{B}\right) \tag{1.2}$$

$$\rho c_p\left[\frac{\partial T}{\partial t} + \vec{V}\cdot\vec{\nabla}T\right] = \vec{\nabla}\cdot k\vec{\nabla}\vec{V} + \rho_P c_{np}\left[D_B\vec{\nabla}C\cdot\vec{\nabla}T + D_T\frac{\vec{\nabla}T\cdot\vec{\nabla}T}{T_\infty}\right] + \mu\Omega \tag{1.3}$$

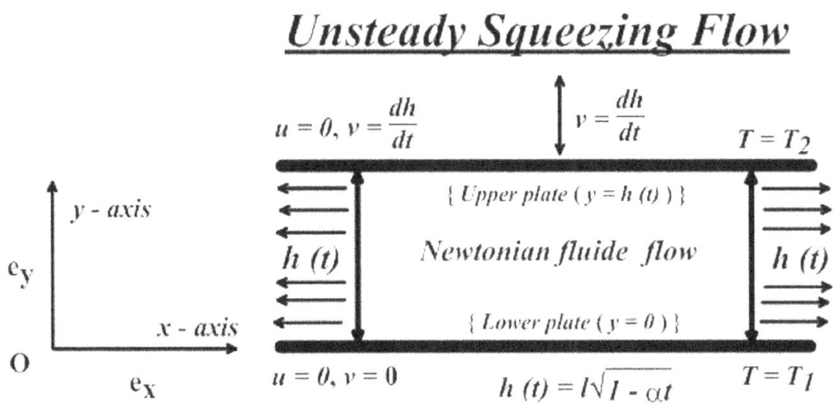

Figure 1.1 Geometry of the present work.

$$\frac{\partial C}{\partial t}+\vec{V}\vec{\nabla}\cdot C=D_B\nabla^2 C+\frac{D_T\cdot\nabla^2 T}{T_\infty} \tag{1.4}$$

In this context, the variables u and v represent the velocities in the x and y directions, respectively. T denotes the temperature, P stands for the pressure, σ signifies the electric conductivity, K denotes the thermal conductivity, and μ represents the dynamic viscosity. In addition, D_B and D_T represent the Brownian motion coefficient and the thermophoretic diffusion coefficient, respectively. The term T refers to the mean fluid temperature.

It is also to establish the operation of $\vec{\nabla}$ as $\vec{\nabla}=\left(\frac{\partial}{\partial x},\frac{\partial}{\partial y},\frac{\partial}{\partial z}\right)$ and Ω. The expression of the viscous dissipation effect is as follows: $\mu\Omega=\mu\left[4\left(\frac{\partial u}{\partial x}\right)^2+\left(\frac{\partial u}{\partial y}+\frac{\partial v}{\partial x}\right)^2\right].$

$$u=0,\quad v=\frac{dh}{dt},\quad T=T_H,\quad C=C_H,\quad \text{at}\quad y=h(t) \tag{1.5}$$

$$v=\frac{du}{dy}=\frac{dT}{dy}=\frac{dC}{dy}=0,\quad \text{at}\quad y=0 \tag{1.6}$$

The following dimensionless groups are introduced:

$$\eta=\frac{y}{\left[l(1-\alpha t)^{\frac{1}{2}}\right]}=\frac{y}{h(t)} \tag{1.7}$$

$$u=\frac{xa}{\left[2(1-\alpha t)\right]}f'(\eta) \tag{1.8}$$

$$v=-\frac{\alpha l}{2(1-\alpha t)^{\frac{1}{2}}}f(\eta) \tag{1.9}$$

$$\Theta=\frac{T}{T_H} \tag{1.10}$$

$$\phi=\frac{C}{C_H} \tag{1.11}$$

By substituting the aforementioned factors into equations (1.2)–(1.4) and eliminating the pressure gradient, the resulting equations can be expressed as follows:

$$f''''+S(ff'''-3f''-\eta f'''-f'f'')-Ha^2 f''=0 \tag{1.12}$$

$$\Theta'' + (P_r N_b)(\Theta' \varphi') + (P_r N_b)\Theta'^2 + (P_r E_b)(f''^2 + 4\delta^2 f'^2) + (P_r S)(f\Theta' - \eta\Theta') = 0 \tag{1.13}$$

$$\varphi'' + (S_c S)(f\varphi' - \eta\varphi') + \left(\frac{N_t}{N_b}\right)\Theta'' = 0 \tag{1.14}$$

The boundary condition transforms into a new set of analogous variables:

$$f''(0) = 0, \quad f(0) = 0, \quad \theta'(0) = 0, \quad \varphi'(0) = 0, \quad \text{at} \quad \eta = 0 \tag{1.15}$$

$$f'(1) = 0, \quad f(1) = 1, \quad \theta(1) = \varphi(1) = 1, \quad \text{at} \quad \eta = 1 \tag{1.16}$$

In this context, the squeeze number (S), Hartmann number (Ha), Brownian motion parameter (Nb), thermophoretic parameter (Nt), Prandtl number (Pr), and Schmidt number (Sc) are defined:

$$Ha = \sqrt{\frac{l^2(1-t)\sigma}{\mu}B_0^2}, \quad S = \frac{\alpha l^2}{2\upsilon}, \quad Pr = \frac{\mu c_p}{K}, \quad Ec = \frac{1}{c_p}\left(\frac{\alpha x}{2(1-\alpha t)}\right)^2, \quad \delta = \frac{l}{x},$$

$$Nb = \frac{\rho c_{pp} D_B C_H}{\rho c_p K}, \quad Nt = \frac{\rho c_{pp} D_T (T_H)}{\left((\rho c_p)_f K\right)}.$$

The skin friction coefficient, Nusselt number, and Sherwood number are the relevant physical quantities of interest, and they are expressed as follows:

$$C_f^* = \frac{\mu}{\rho v_w^2} \frac{\partial u}{\partial y}\bigg|_{y=h(t)} \tag{1.17}$$

$$Nu^* = -\left(\frac{l(k)\dfrac{\partial T}{\partial y}\bigg|_{y=h(t)}}{KT_H}\right) \tag{1.18}$$

$$Sh^* = -\left(\frac{l(D)\dfrac{\partial C}{\partial y}\bigg|_{y=h(t)}}{DC_H}\right) \tag{1.19}$$

Using variables (1.7–1.11), we get

$$Nu = -\theta'(1) \tag{1.20}$$

$$C_f = -f''(1) \tag{1.21}$$

$$Sh = -\varphi'(1) \tag{1.22}$$

1.3 THE HOMOTOPY PERTURBATION METHOD

When addressing the system of nonlinear differential equations outlined in equations (1.12)–(1.14), a thorough evaluation of the chosen method becomes crucial. The selected method is favored for its distinct advantages, as it has displayed satisfactory analytical results with desirable convergence and stability (Dogonchi & Ganji, 2016; He, 1999). By integrating this method with the comprehensive analytical procedures of the HPM, the fundamental principles can be effectively illustrated. To exemplify this, let us consider the following equation:

$$A(u) - f(r) = 0, \quad r \in \Omega \tag{1.23}$$

With the boundary condition of

$$B\left(u, \frac{\partial U}{\partial n}\right) = 0, \quad r \in \Gamma \tag{1.24}$$

In this context, A represents a general differential operator, B denotes a boundary operator, $f(r)$ is a known analytical function within the domain Ω, and Γ represents the boundary of the domain.

A can be partitioned into two parts: L and N. Specifically, L is linear, whereas N is nonlinear in nature.

Equation (1.23) can, therefore, be rewritten as

$$L(u) + N(u) - f(r) = 0 \tag{1.25}$$

Homotopy perturbation structure is shown as follows:

$$H(v,p) = (1-p)\left[L(v) - L(u_0)\right] + p\left[L(v) + N(v) - f(r)\right] = 0 \tag{1.26}$$

where

$$v(r,p) : \Omega \times [0,1] \rightarrow R \tag{1.27}$$

In equation (1.26), the parameter p, which lies in the range [0, 1], serves as an embedding parameter, while v_o represents the first approximation that satisfies the boundary condition. It is assumed that the solution to equation (1.23) can be expressed as a power series in p:

$$v = v_0 + pv_1 + p^2v_2 + p^3v_3 + \cdots \tag{1.28}$$

The optimal approximation for the solution is obtained as

$$u = \lim_{p\to 1} v = v_0 + v_1 + v_2 + \cdots \tag{1.29}$$

By applying the HPM to address the problem, the following expression is obtained:

$$H(f,p) = (1-p)\left[f'''' - f_0''''(0)\right] + p\left[f'''' + S(ff''' - 3f'' - \eta f''' - f'f'') - Ha^2 f'' = 0\right] \tag{1.30}$$

$$\begin{aligned} H(\theta,p) = (1-p)\left[\theta'' - \theta_0''(0)\right] + p\Big[\theta'' + (P_rN_b)(\Theta'\varphi') + (P_rN_t)\Theta'^2 \\ + (P_rE_c)(f''^2 + 4\delta^2 f'^2)(-\eta f''' - f'f'') + (P_rS)(f\Theta' - \eta\Theta') = 0\Big] \end{aligned} \tag{1.31}$$

f, θ, and φ are considered as follows:

$$f(\eta) = f_0(\eta) + p^1 f_1(\eta) + p^2 f_2(\eta) + p^3 f_3(\eta) + \cdots = \sum_{i=0}^{N} p^i f_i(\eta) \tag{1.32}$$

$$\theta(\eta) = \theta_0(\eta) + p^1\theta_1(\eta) + p^2\theta_2(\eta) + p^3\theta_3(\eta) + \cdots = \sum_{i=0}^{N} p^i\theta_i(\eta) \tag{1.33}$$

$$\varphi(\eta) = \varphi_0(\eta) + p^1\varphi_1(\eta) + p^2\varphi_2(\eta) + p^3\varphi_3(\eta) + \cdots = \sum_{i=0}^{N} p^i\varphi_i(\eta) \tag{1.34}$$

By substituting the values of f, θ, and φ from equations (1.33)–(1.35) into equations (1.30)–(1.32) and performing simplifications and rearrangements based on the powers of p, the resulting expressions can be obtained:

p^0 :

$$f_0'''' = 0 \tag{1.35}$$

$$\theta_0'' = 0 \tag{1.36}$$

$$\varphi_0'' = 0 \tag{1.37}$$

And boundary conditions are

$$f''(0)=0,\quad f(0)=0,\quad \theta'(0)=0,\quad \phi'(0)=0,\quad \text{at} \quad \eta=0 \tag{1.38}$$

$$f'(1)=0,\quad f(1)=1,\quad \theta(1)=\phi(1)=1,\quad \text{at} \quad \eta=1 \tag{1.39}$$

p^1:

$$f_1'''' + S(f_0 f_0''' - 3f_0'' - \eta f_0''' - f_0' f_0'') - Ha^2 f_0'' = 0 \tag{1.40}$$

$$\theta_1'' + (P_r E_c)(f_0''^2 + 4\delta^2 f_0'^2) + (P_r S)(f_0 \theta_0' - \eta \theta_0') + (P_r N_b)(\theta_0' \varphi_0) + (P_r N_t)\theta_0'^2 = 0 \tag{1.41}$$

$$\varphi_1'' + (S_C S)(f_0 \varphi_0' - \eta \varphi_0') + \left(\frac{Nt}{Nb}\right)\theta_0'' \tag{1.42}$$

Boundary conditions are

$$f_1''(0)=0,\quad f_1(0)=0,\quad \theta_1'(0)=0,\quad \varphi_1'(0)=0 \quad at \quad \eta=0 \tag{1.43}$$

$$f_1'(1)=0,\quad f_1(1)=0,\quad \theta_1(1)=0,\quad \varphi_1(1)=0 \quad \text{at} \quad \eta=1 \tag{1.44}$$

By solving equations (1.29) and (1.31) with the specified boundary conditions, the following solutions are obtained:

$$f_0(\eta) = -\frac{1}{2}\eta^3 + \frac{3}{2}\eta \tag{1.45}$$

$$\theta_0(\eta) = 1 \tag{1.46}$$

$$\varphi_0(\eta) = 1 \tag{1.47}$$

$$f_0(\eta) = \frac{1}{280}S\eta^7 - \frac{1}{40}\left(Ha^2 + 4S\right)\eta^5 + \frac{1}{6}\left(\frac{3}{10}Ha^2 + \frac{159}{140}S\right)\eta^3 + \left(-\frac{1}{40}Ha^2 - \frac{13}{140}S\right)\eta \tag{1.48}$$

$$\theta_1(\eta) = -9P_r E_c\left(\frac{1}{30}\delta^2\eta^6 - \frac{1}{6}\delta^2\eta^4 + \frac{1}{12}\eta^4 + \frac{1}{2}\delta^2\eta^2\right) + \frac{33}{10}E_c P_r \delta^2 + \frac{3}{4}P_r E_c \tag{1.49}$$

$$\varphi_1(\eta) = 1 \tag{1.50}$$

The terms $f_i(\eta)$, $\theta_i(\eta)$ and $\varphi_i(\eta)$ when $i > 2$ are too large that is mentioned graphically.

The solution to the equations, as $p \to 1$, is given by

$$f(\eta) = f_0(\eta) + p^1 f_1(\eta) + p^2 f_2(\eta) + p^3 f_3(\eta) + \cdots = \sum_{i=0}^{N} p^i f_i(\eta) \tag{1.51}$$

$$\theta(\eta) = \theta_0(\eta) + p^1\theta_1(\eta) + p^2\theta_2(\eta) + p^3\theta_3(\eta) + \cdots = \sum_{i=0}^{N} p^i \theta_i(\eta) \tag{1.52}$$

$$\varphi(\eta) = \varphi_0(\eta) + p^1\varphi_1(\eta) + p^2\varphi_2(\eta) + p^3\varphi_3(\eta) + \cdots = \sum_{i=0}^{N} p^i \varphi_i(\eta) \tag{1.53}$$

1.4 BASIC IDEA OF AGM

The procedure has been comprehensively explained by using straightforward language to enhance the understanding of the method presented in this chapter. Adhering to the boundary conditions, the general form of a differential equation can be expressed as follows:

$$p_k : f(u, u', u'', \ldots, u^{m}) = 0; \quad u = u(x) \tag{1.54}$$

The nonlinear differential equation of p is considered, where p is a function of u and u is a function of x along with their respective derivatives as follows:

Boundary conditions:

$$\begin{aligned} &u(x) = u_0, \quad u'(x) = u_1, \ldots, u^{(m-1)}(x) = u_{m-1}; \quad x = 0 \\ &u(x) = u_{L_0}, \quad u'(x) = u_{L_1}, \ldots, u^{(m-1)}(x) = u_{L_{m-1}}; \quad x = L \end{aligned} \tag{1.55}$$

To solve the first-order differential equation with the boundary conditions at $x = L$ as described in equation (1.68), an assumed solution in the form of a sequence of letters with constant coefficients is employed. This sequence is represented by the expression:

$$u(x) = \sum_{i=0}^{n} a_i x^i = a_0 + a_1 x^1 + a_2 x^2 + a_3 x^3 + \cdots + a_n x^n \tag{1.56}$$

Equation (1.54) provides more options for series representations, allowing for a more accurate solution. In this case, there are $(n+1)$ unknown coefficients that can be determined by specifying $(n+1)$ equations to find the solution for equation (1.55) corresponding to the series up to degree (n). The set of $(n+1)$ equations is then solved using the boundary conditions given by equation (1.55).

1.4.1 Boundary conditions implementation

Implementation of boundary conditions for differential equation (1.56):

When $x = 0$:

$$\begin{aligned} u(0) &= a_0 = u_0 \\ u'(0) &= a_1 = u_1 \\ u''(0) &= a_2 = u_2 \end{aligned} \tag{1.57}$$

And when $x = L$:

$$\begin{aligned} u(L) &= a_0 + a_1 L + a_2 L^2 + \cdots + a_n L^n = u_{(L_0)} \\ u'(L) &= a_1 + 2a_2 L + 3a_3 L^2 + \cdots + n a_n L^{(n-1)} = u_{(L_1)} \\ u''(L) &= 2a_2 + 6a_3 L + 12a_4 L^2 + \cdots + n(n-1) a_n L^{(n-2)} = u_{(L_2)} \end{aligned} \tag{1.58}$$

Substituting equation (1.58) into equation (1.54), the subsequent step involves incorporating the boundary conditions into the differential equation (1.54) using the following procedure:

$$\begin{aligned} p_0 &: f\left(u(0), u'(0), u''(0) \ldots\ldots u^{(m)}(0)\right) \\ p_1 &: f\left(u(L), u'(L), u''(L) \ldots\ldots u^{(m)}(L)\right) \end{aligned} \tag{1.59}$$

In terms of selecting n, a subset of equations ($n<m$) from equation (1.56) is considered, forming a set of ($n+1$) equations with ($n+1$) unknowns. However, this set includes additional unknowns that are essentially the same coefficients mentioned in equation (1.56). To determine these additional unknowns using the boundary conditions, it is necessary to deduce m equations from equation (1.54) a total of m times.

$$\begin{aligned} p'_k &: f\left(u', u'', u''', \ldots, u^{(m+1)}\right) \\ p''_k &: f\left(u', u'', u^{(IV)}, \ldots, u^{(m+2)}\right) \end{aligned} \tag{1.60}$$

To handle the boundary conditions concerning the derivatives of the differential equation P_k in equation (1.60), they are addressed in the following manner:

p_k' :

$$f\left(u'(0),u''(0),u'''(0),\ldots,u^{(m+1)}(0)\right) \tag{1.61}$$

$$f\left(u'(L),u''(l),u'''(L),\ldots,u^{(m+1)}(L)\right)$$

p_k'' :

$$f\left(u'(0),u''(0),u^{(IV)}(0),\ldots,u^{(m+2)}(0)\right) \tag{1.62}$$

$$f\left(u''(L),u'''(l),u^{(IV)}(L),\ldots,u^{(m+1)}(L)\right)$$

To determine the $(n+1)$ unknown coefficients (a_0, a_1, a_2, ..., a_n) of equation (1.56), a set of $(n+1)$ equations can be derived from equation (1.57) to equation (1.62). Solving this set of equations allows us to calculate the coefficients required to solve the nonlinear differential equation (1.54). By determining the coefficients specified in equation (1.56), the solution to equation (1.54) can be obtained. To enhance comprehension of the process, a step-by-step procedure is outlined in the subsequent section.

1.4.2 Application of Akbari–Ganji's Method (AGM)

Considering the given system of coupled nonlinear differential equations and bearing in mind the fundamental concept of the method, equations (1.12), (1.13), and (1.14) are rearranged in the following order:

$$F(\eta) = f'''' + S(ff''' - 3f'' - \eta f''' - f'f'') - Ha^2 f'' = 0 \tag{1.63}$$

$$\begin{aligned} G(\eta) &= \theta'' + (P_r N_b)(\theta'\varphi') + (P_r N_t)\theta'^2 \\ &\quad + (P_r E_c)(f''^2 + 4^2\delta f'^2) + (P_r S)(f\theta' - \eta\theta') = 0 \end{aligned} \tag{1.64}$$

$$\varphi'' + (S_c S)(f\varphi' - \eta\varphi') + \left(\frac{N_t}{N_b}\right)\theta'' = 0 \tag{1.65}$$

Following the fundamental concept of AGM, a suitable trial function is employed as the solution for the given differential equation. To represent this trial function, it is expressed as a finite series of polynomials with constant coefficients, as described below:

$$\begin{aligned} f(\eta) &= \sum_{i=0}^{9} a_i\eta^i = a_0 + a_1\eta^1 + a_2\eta^2 + a_3\eta^3 + a_4x^4 + a_5x^5 \\ &\quad + a_6x^6 + a_7x^7 + a_8x^8 + a_9x^9 \end{aligned} \tag{1.66}$$

$$\theta(\eta) = \sum_{i=0}^{9} b_i\eta^i = b_0 + b_1\eta^1 + b_2\eta^2 + b_3\eta^3 + b_4x^4 + b_5x^5 + b_6x^6 + b_7x^7 + b_8x^8 + b_9x^9 \tag{1.67}$$

$$\varphi(\eta) = \sum_{i=0}^{9} b_i\eta^i = b_0 + b_1\eta^1 + b_2\eta^2 + b_3\eta^3 + b_4x^4 + b_5x^5 + b_6x^6 + b_7x^7 + b_8x^8 + b_9x^9 \tag{1.68}$$

1.4.3 Results and discussion

The primary objective of this study is to address the issue of magnetohydrodynamic squeezing flow between parallel plates using two distinct methods: the HPM and the AGM. The investigation evaluates the influence of various significant parameters on the heat and mass criteria. The results presented in Table 1.1 indicate that the proposed approach yields highly accurate solutions for this problem. To validate the effectiveness of the implemented code, the obtained outcomes are compared with those of other referenced studies (Dogonchi & Ganji, 2016; Mustafa et al., 2012). The findings demonstrate a high level of consistency and agreement.

Figure 1.2 illustrates the variations in the velocity profile with changing Hartmann number (Ha) and squeeze number (S). An increase in the Hartmann number induces a Lorentz force in the perpendicular direction, resulting in a reduction in the velocity profile. In addition, as the plates compress the flow, leading to a higher squeeze number, the kinematic viscosity decreases, further diminishing the velocity.

Table 1.1 Comparison between the present work and Mustafa et al.'s and Dogonchi et al.'s results for $-f''(1)$ and $-\theta'(1)$ when $Ec = Pr = 1$, $Sc = Nb = Nt = 0$, $Ha = 0$, $\delta = 0.1$ for various numbers of S

	Mustafa et al. (2012)		*Dogonchi and Ganji (2016)*		*Present results ($Ha = 0, Nb = Nt = Sc = 0$)*			
	HAM		*DRA*		*HPM*		*AGM*	
S	$-f''(1)$	$-\theta'(1)$	$-f''(1)$	$-\theta'(1)$	$-f''(1)$	$-\theta'(1)$	$-f''(1)$	$-\theta'(1)$
−1.0	2.170090	3.319899	2.170091	3.319888	2.170092	3.319860	2.170090	3.319899
−0.5	2.614038	3.129491	2.617403	3.129491	2.617403	3.129491	2.617403	3.129491
0.01	3.007134	3.047092	3.007133	3.047091	3.007133	3.047091	3.007133	3.047091
0.5	3.336449	3.026324	3.336449	3.026323	3.336449	3.026323	3.336449	3.026323
2.0	4.167389	3.118551	4.167041	3.113386	4.168065	3.127819	4.167389	3.118550

During the computation, the following parameter values are considered: $S = 1$, $Ha = 1$, $Pr = 6.2$, $Ec = 0.01$, $\delta = 0.1$, $Nb = 0.1$, $Nt = 0.1$, $Sc = 1$.

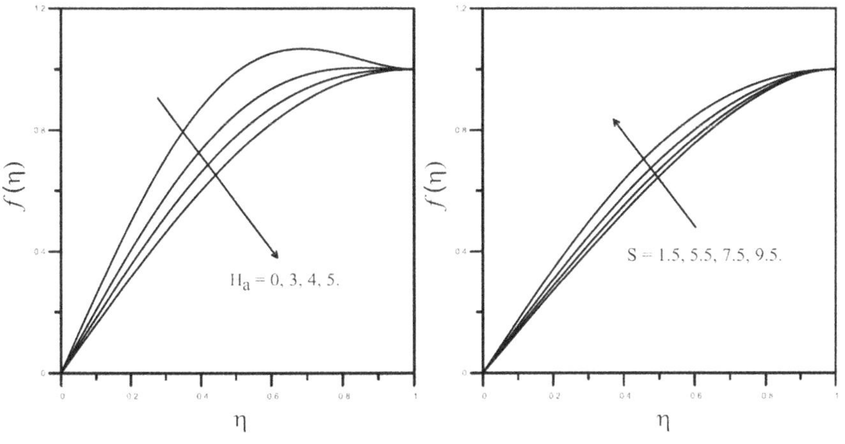

Figure 1.2 Effect of the Hartman and squeeze numbers on the velocity.

Figure 1.3 displays the correlation between the Hartmann number and the squeeze number concerning the temperature profile. An increase in the Hartmann number corresponds to a decrease in the temperature profile. Similarly, an increase in the squeeze number leads to a reduction in the temperature profile.

Figure 1.4 presents the influence of the Schmidt number (*Sc*) on the concentration profile and the impact of the Eckert number (*Ec*) on temperature. The Eckert number accounts for the self-heating effect due to dissipation within the fluid. The graph demonstrates that as the Eckert number increases, the temperature profile rises. This phenomenon occurs due to the combined effect of temperature gradients and dissipation caused

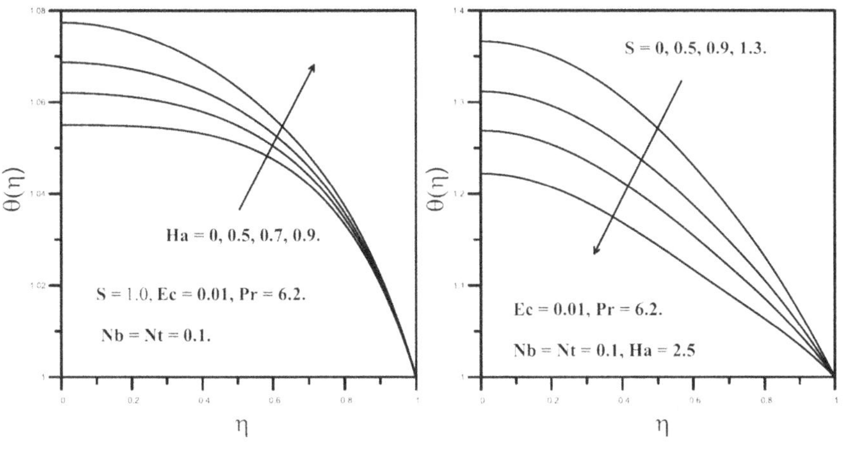

Figure 1.3 Effect of the Hartman and squeeze numbers on the temperature.

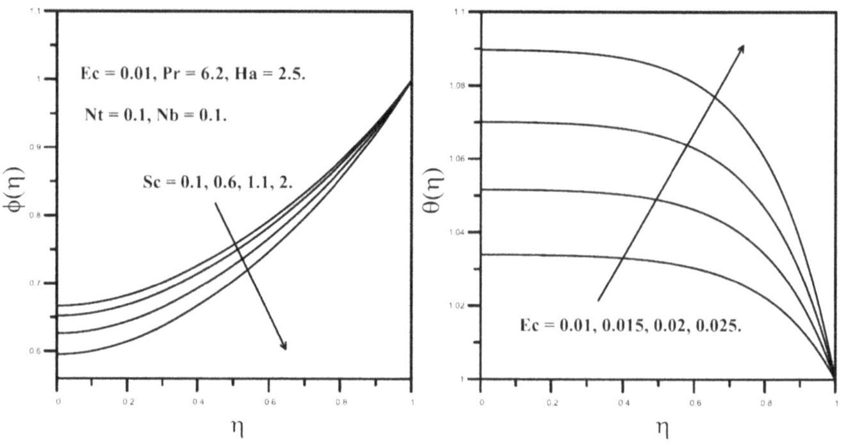

Figure 1.4 Effect of the Eckert number on the temperature and the effect of the Schmidt number on concentration.

by internal friction within the nanofluid. On the other hand, the Schmidt number's effect on concentration reveals that an increase in the Schmidt number results in an elevated concentration.

Figure 1.5 depicts the influence of the thermophoretic parameter (Nt) and the Brownian motion factor (Nb) on the temperature profile. The findings show that as the thermophoretic parameter increases, the temperature profile experiences an elevation. However, variations in the Brownian motion parameters have a negligible effect on the temperature profile.

Figure 1.6 demonstrates the effects of the thermophoretic parameter (Nt) and the Brownian motion parameter (Nb) on the concentration profile.

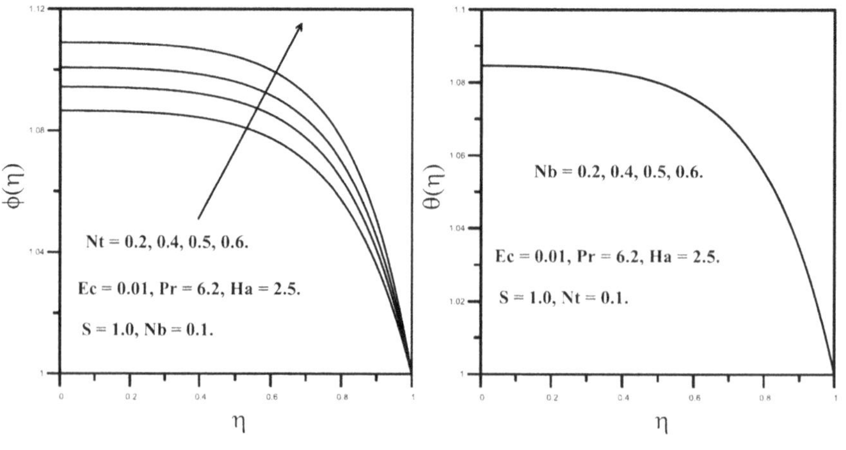

Figure 1.5 Effect of thermophoretic and Brownian motion parameters on temperature.

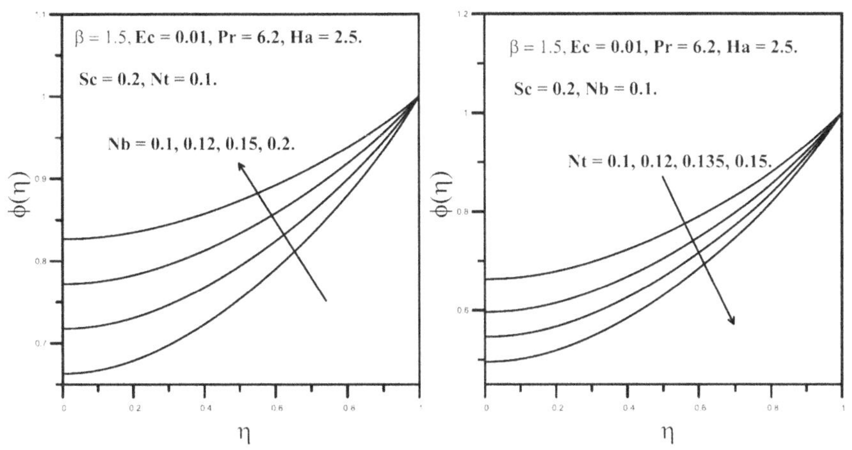

Figure 1.6 Effect of thermophoretic and Brownian motion parameters on concentration.

The graph reveals that both parameters exert a similar influence on the concentration profile. As the values of Nt and Nb increase, the concentration also increases.

Figure 1.6 illustrates the impact of the thermophoretic parameter (Nt) and the Brownian motion parameter (Nb) on the concentration profile. The plot shows that both parameters have a similar effect on the concentration profile, where higher values of Nt and Nb lead to an increased concentration.

In addition, it is worth noting that as the Hartmann number increases, there is a simultaneous decrease observed in the skin friction coefficient (Cf), Nusselt number (Nu), and Sherwood number (Sh). This inverse correlation suggests that an increase in the Hartmann number results in a reduction in these coefficients (Tables 1.2 and 1.3).

Table 1.2 Variation of C_f (skin friction coefficient), Nu (Nusselt number) and Sh (Sherwood number) with different values of Squeeze number S

S	$C_f\left(f''(1)\right)$	$Nu\left(-\theta'(1)\right)$	$Sh\left(-\varphi'(1)\right)$
0.0	–3.19452804947	0.189277021948	–0.189268877588
0.1	–3.26091278008	0.186574307426	–0.186082155965
0.3	–3.3892861993	0.182214705142	–0.180875287166
0.5	–3.51230505348	0.17899803036	–0.176886220495
0.7	–3.63049919142	0.17672408763	–0.173749600961
0.9	–3.74432481098	0.175322951881	–0.171075308593
1.0	–3.79972587711	0.174975748129	–0.169757993785
1.1	–3.85418093078	0.174909570311	–0.168343798811

Table 1.3 Variation of C_f (skin friction coefficient), Nu (Nusselt number), and Sh (Sherwood number) with different values of Hartman number Ha

Ha	$c_f\left(f''(1)\right)$	$Nu\left(-\theta'(1)\right)$	$Sh\left(-\varphi'(1)\right)$
0.0	−3.63839511718	0.17141127277	−0.167029947093
0.1	−3.64004184376	0.171445453724	−0.16705902804
0.3	−3.65319058258	0.17171981908	−0.16729061397
0.5	−3.67935548763	0.17227363022	−0.167747791552
0.7	−3.71827686546	0.173118027667	−0.168416885747
0.9	−3.76957841861	0.174272602644	−0.169272711796
1.0	−3.79972587711	0.174975748129	−0.169757993785
1.1	−3.83278216446	0.175770215727	−0.170271534667

1.5 CONCLUSION

The present research employs the HPM to thoroughly investigate the dynamics of unstable squeezing nanofluid flow between parallel plates under the influence of a varying magnetic field. The main focus is on analyzing the impact of key parameters, including the squeezing number, Hartmann number, Eckert number, Brownian motion parameter, thermophoretic parameter, and Schmidt number, to gain a comprehensive understanding of their effects on flow behavior and characteristics.

The obtained findings from this investigation strongly align with previous works, confirming the effectiveness of the HPM in addressing complex problems of this nature. Specifically, an increase in the thermophoretic parameter (Nt) leads to higher temperature values and lower nanoparticle concentration, whereas an increase in the Hartmann number results in an elevation of the temperature profile and a reduction in the velocity profile. Furthermore, various other aspects have been explored, and relevant results have been presented.

Overall, this study highlights the practical applicability of the HPM for analyzing unstable squeezing nanofluid flow between parallel plates and offers valuable insights into the effects of different parameters on the system behavior. By shedding light on these phenomena, the research contributes to the existing knowledge in this field, paving the way for further exploration and potential engineering applications.

NOMENCLATURE

θ	Dimensionless temperature
μ	Dynamic viscosity of nanofluid
ρ	Density

σ	Electrical conductivity of nanofluid
φ	Dimensionless concentration
B	Magnetic field
C	Nanofluid concentration
cp	Specific heat at constant pressure
DB	Brownian diffusion coefficient
DT	Thermophoretic diffusion coefficient
Ec	Eckert number
Ha	Hartmann number
k	Thermal conductivity
l	Distance of plate
Nt	Thermophoretic parameter
p	Pressure
S	Squeeze number
T	Fluid temperature
Nb	Brownian motion parameter
Nu	Nusselt number
Pr	Prandtl number
Sc	Schmidt number

REFERENCES

Dogonchi, A. S., & Ganji, D. D. (2016). Investigation of MHD nanofluid flow and heat transfer in a stretching/shrinking convergent/divergent channel considering thermal radiation. *Journal of Molecular Liquids*, *220*, 592–603.

Domairry, G., & Aziz, A. (2009). Approximate analysis of MHD squeeze flow between two parallel disks with suction or injection by homotopy perturbation method. *Mathematical Problems in Engineering*, *2009*, 603916.

Draper, B., Yee, W. L., Pedrana, A., Kyi, K. P., Qureshi, H., Htay, H., Naing, W., Thompson, A. J., Hellard, M., & Howell, J. (2022). Reducing liver disease-related deaths in the Asia-Pacific: the important role of decentralised and non-specialist led hepatitis C treatment for cirrhotic patients. *The Lancet Regional Health-Western Pacific*, *20*, 100359.

El Harfouf, A., Hayani Mounir, S., & Wakif, A. (2023). Steady magnetohydrodynamic Casson nanofluid flow between two infinit parallel plates using Akbari Ganji's Method (AGM). *Journal of Nanofluids*, *12*(3), 633–642.

Hayat, T., Yousaf, A., Mustafa, M., & Obaidat, S. (2012). MHD squeezing flow of second-grade fluid between two parallel disks. *International Journal for Numerical Methods in Fluids*, *69*(2), 399–410.

He, J.-H. (1999). Homotopy perturbation technique. *Computer Methods in Applied Mechanics and Engineering*, *178*(3–4), 257–262.

He, J.-H. (2004). Comparison of homotopy perturbation method and homotopy analysis method. *Applied Mathematics and Computation*, *156*(2), 527–539.

Jang, M.-J., Chen, C.-L., & Liu, Y.-C. (2001). Two-dimensional differential transform for partial differential equations. *Applied Mathematics and Computation*, *121*(2–3), 261–270.

Mahmood, M., Asghar, S., & Hossain, M. A. (2007). Squeezed flow and heat transfer over a porous surface for viscous fluid. *Heat and Mass Transfer*, *44*, 165–173.

Mirgolbabaee, H., Ledari, S. T., & Ganji, D. D. (2016). New approach method for solving Duffing-type nonlinear oscillator. *Alexandria Engineering Journal*, *55*(2), 1695–1702.

Mohyud-Din, S. T., & Yildirim, A. (2010). Ma's variation of parameters method for Fisher's equations. *Advances in Applied Mathematics and Mechanics*, *2*(3), 379–388.

Mustafa, M., Hayat, T., & Obaidat, S. (2012). On heat and mass transfer in the unsteady squeezing flow between parallel plates. *Meccanica*, *47*, 1581–1589.

Sheikholeslami, M., Abelman, S., & Ganji, D. D. (2014a). Numerical simulation of MHD nanofluid flow and heat transfer considering viscous dissipation. *International Journal of Heat and Mass Transfer*, *79*, 212–222.

Sheikholeslami, M., Azimi, M., & Ganji, D. D. (2015). Application of differential transformation method for nanofluid flow in a semi-permeable channel considering magnetic field effect. *International Journal for Computational Methods in Engineering Science and Mechanics*, *16*(4), 246–255.

Sheikholeslami, M., Ellahi, R., Ashorynejad, H. R., Domairry, G., & Hayat, T. (2014b). Effects of heat transfer in flow of nanofluids over a permeable stretching wall in a porous medium. *Journal of Computational and Theoretical Nanoscience*, *11*(2), 486–496.

Sheikholeslami, M., & Ganji, D. D. (2015). Nanofluid flow and heat transfer between parallel plates considering Brownian motion using DTM. *Computer Methods in Applied Mechanics and Engineering*, *283*, 651–663.

Sheikholeslami, M., & Ganji, D. D. (2016). Nanofluid hydrothermal behavior in existence of Lorentz forces considering Joule heating effect. *Journal of Molecular Liquids*, *224*, 526–537.

Sheikholeslami, M., & Ganji, D. D. (2017). Impact of electric field on nanofluid forced convection heat transfer with considering variable properties. *Journal of Molecular Liquids*, *229*, 566–573.

Sheikholeslami, M., Ziabakhsh, Z., & Ganji, D. D. (2017). Transport of magnetohydrodynamic nanofluid in a porous media. *Colloids and Surfaces A: Physicochemical and Engineering Aspects*, *520*, 201–212.

Siddiqui, A. M., Irum, S., & Ansari, A. R. (2008). Unsteady squeezing flow of a viscous MHD fluid between parallel plates, a solution using the homotopy perturbation method. *Mathematical Modelling and Analysis*, *13*(4), 565–576.

Stefan, J. (1875). Versuche über die scheinbare Adhäsion. *Annalen Der Physik*, *230*(2), 316–318.

Chapter 2

Underground coal gasification modelling

Devansh Shrivastava
University of Aberdeen

2.1 INTRODUCTION

Underground coal gasification (UCG) is one of the best and most promising options for the processing of unused coal in the future. As the world is moving towards unconventional sources of energy, this process enhances this change towards the unconventional by allowing coal to be gasified in situ within the coal bed with the help of a series of interconnected wells. The coal is ignited with specialised techniques and the air is flown in to sustain the fire being ignited, which is essentially required to reach the coal and produce a combustible synthetic gas further useful for industrial processes such as the manufacturing of hydrogen or diesel fuel or gas or most importantly power generation.

Field tests based on UCG have been in practice since the 1930s. These tests have seen both success and failures, and since then, the number of tests has seen an increase, but the plants in functioning are very few due to the varying degrees of success and the fluctuating prices of oil and gas (Perkins, 2018; Magnani and Farouq Ali, 1975); based on this experience (especially in the USSR and the United States), field designs that apply to a varied classification of geological conditions and coal properties have been developed and brought to use. In the past, countries that have a significant amount of low-rank coal reserves increased their attention on the UCG process and related research activities (Perkins et al., 2016; Mocek et al., 2016; Laciak et al., 2016; Nourozieh et al., 2010).

Appropriate field and laboratory results as well as pre-designed mathematical models of an in situ gasifier are observed and tested as part of the analysis. Many UCG tests were conducted on-site and in labs whether in situ or ex-situ, and they showed somewhat similar results. UCG tests in China proved facts showing that when steam flows into a coal bed, syngas can be obtained containing more than 50% hydrogen concentration (Li, Liang and Liang, 2007). These tests incorporated a two-step gasification process, wherein the first step was to supply O_2-rich air into the field to maintain combustion and the second step was to replace the O_2-rich air with steam to start the water gasification reaction to produce hydrogen (Yang et al., 2008).

DOI: 10.1201/9781032712079-2

Yang et al. increased the hydrogen production rate of the two-step gasification process by providing the gasification agents through multiple locations, which was eventually a form of multilateral formation (Yang et al., 2008).

The estimated reserves of crude oil and natural gas in India as of 31.03.2018 stood at 594.49 million metric tonnes (MMT) and 1339.57 billion cubic metres (BCM), respectively. As of 31.03.18, the estimated reserves of coal were around 319.04 billion tones, an addition of 3.88 billion over the last year. Coal deposits are mainly confined to the eastern and south-central parts of the country.

There has been an increase of 1.23% in the estimated coal reserves during the year 2017–2018 with Odisha accounting for the maximum increase of 2.6%, whereas there was a decrease of 1.59% in the estimated reserve of crude oil for the country as a whole during 2017–2018 as compared with the position a year ago.

Hence, UCG is the only economically profitable method to harness the inaccessible reserves. There are a few organisations looking to set up UCG projects in Gujarat and Rajasthan. A pilot project of UCG was conducted by ONGC in collaboration with Gujarat Industries Power Company Ltd in Surat, Gujarat. Now, as the process of UCG has established its significance in the field of unconventional sources, the organisation has taken over the Vastan mines in Surat and in collaboration with Messrs. National Mining Research Center-Skochinsky Institute of Mining (NMRC-SIM), Russia is progressing towards their goal to complete this project.

In situ gasification of coal exposes the groundwater to a potential environmental hazard that is mainly a result of local hydrogeological conditions but is somewhat affected by the process of UCG. Further analysis of the samples brings out the effects of UCG on the groundwater. The process of UCG is a complex procedure with a series of techniques involved in it. These techniques often lead to the production of unwanted traces of pollutants that can be transmitted to the surrounding strata by the processes of diffusion or direct injection. Most commonly found pollutants in the groundwater strata may be divided into categories as minor, organic and inorganic. A mathematical model is often useful to attribute the flow of groundwater once the process of UCG takes place and this can prove the effects of contaminants. Some of the organic contaminants are phenols, benzenes, naphthalene, toluene and xylene. Some of the inorganic contaminants are ammonium, boron, calcium, iron, lead, magnesium, manganese, zinc, mercury and sulphate.

The shift from a conventional process to an unconventional one can never be achieved with ease. However, once in use, it can be an alternative to the orthodox methods. UCG is one such example, and the aim of this study was to apprise readers with the process and its underlying dynamics. Theoretically defined, UCG is an industrial process that aims towards the in situ combustion of a coalseam leading to the production of product gases that can be extracted for various purposes.

Apart from these ex-situ and in situ processes, several studies incorporate mathematical modelling which also employs computational fluid dynamics (CFD). Perkins and Sahajwalla created a 2D Geometry to simulate and calculate the mass transfer and heat transfer operations inside the seam during the process of UCG (Perkins and Sahajwalla, 2007). Turbulence models were used to investigate and model the effects of turbulent flow due to natural convection and had a low value of K-Epsilon. Sarraf Shirazi et al. created a 3D model in CFD that consisted of a coal seam and a gasification chamber (Sarraf Shirazi, Karimipour and Gupta, 2013). Using the ANSYS Fluent simulator, numerical results were obtained. Żogała and Janoszek have also performed some notable work in the field of UCG with the help of a 3D model in CFD which had a design similar to a gas reactor (Żogała and Janoszek, 2015). Using ANSYS Fluent solver (Coffield and Shepherd, 1987), the authors calculated the effects of the flow of steam on the temperature levels inside the reactor and the produced syngas composition.

In this study, a 2D mathematical model that was dynamic in order and based on kinetic reaction rates was created and investigated in ANSYS Fluent to evaluate numerical results. The mathematical model prepared in the GeometryModeler of the ANSYS Workbench consists of the inlet and outlet streams and a coal reactor part where the reactions take place. Table 2.1 shows the various enthalpies for the listed reactions. There is a total of six species involved in this work mentioned as O_2, CO_2, CO, H_2O, H_2 and N_2. For these six chemical species, there are six chemical reactions involved (reactions 2.3–2.8) apart from the reactions of drying and pyrolysis (reactions 2.1 and 2.2).

Syngas stands for synthetic gas, which is a combination of gases such as hydrogen and carbon monoxide prominently; however, there are traces of carbon dioxide found as well. It is equivalent to almost half the energy density of natural gas, but it cannot be used directly as a fuel and hence is used to establish fuel sources. Syngas is a primitive part of the chemical industry, which relates to almost 2% of the total primary energy usage. Syngas obtained is not completely free of impurities and contains some traces which are further eliminated via processing. Syngas can be derived

Table 2.1 Various enthalpies for the listed reactions

Process	*Reaction*	*ΔH (kJ/mol)*
Drying	(2.1) Coal → Coal (dry) + H_2O	+40
Pyrolysis	(2.2) Coal (dry) → Volatile matter + Char	0
CO combustion	(2.3) $CO + ½O_2 \rightarrow CO_2$	−111
H_2 combustion	(2.4) $H_2 + ½O_2 \rightarrow H_2O$	−242
Water–gas shift reaction	(2.5) $CO + H_2O \rightleftarrows CO_2 + H_2$	−41
Combustion	(2.6) $C + O_2 \rightarrow CO_2$	−393
CO_2 gasification	(2.7) $C + CO_2 \rightarrow 2CO$	+172
H_2O gasification	(2.8) $C + H_2O \rightarrow CO + H_2$	+131

as a product of gasification from several sources such as biomass, coal and natural gas, by reaction with steam thermally breaking down the biomass into a combustible gas in a closed reactor is known as thermal gasification and it produces by-products such as volatiles, char and ash as well.

In situ, combustion can be defined as a method of recovery of fuel in which fire is generated inside the reservoir by injecting air containing oxygen. In situ combustion is also known as the process of fire flooding and is one of the oldest and most used methods of recovering oil using tertiary methods, from the reservoir. Specialised heaters in the well are accommodated to ignite the oil in the reservoir and start a fire. A continuous flow of air is essential to keep the fire ignited and hence air or oxygen-rich air is made to flow through the well (Bhutto, Bazmi and Zahedi, 2013). This flow of air or the entire process of fire flooding is dependent on the direction of the flow or propagation and is further classified as Forward or Reverse combustion. The high temperatures in the well lead to hydrocarbon cracking and the vaporisation of light hydrocarbons. The process of breaking long-chain hydrocarbons into lesser complex products has been essential and widely used in the petroleum industry.

2.2 THE ANALYTICAL STUDY

A two-dimensional (2D) mathematical model along with its solution geometry having the dimensions of a UCG reactor was developed using the ANSYS Workbench and ANSYS Fluent software environment. Developing a mathematical model in two dimensions and working on it for its solution geometry is less complex and sufficient to analyse the effects of gasification parameters.

A 2D model is first created using the ANSYS Workbench and then a grid is created using ANSYS Meshing. The CFD model created using the software environment is used to investigate the thermal and chemical heat and mass transfer interactions between the source and parameters.

2.3 EQUATIONS INVOLVED IN THE STUDY (ANDERSON, 1995)

- Continuity Equation:

$$\frac{\partial \rho}{\partial t} + \text{div}(\rho \vec{u}) = s_m$$

- Momentum Equation:

$$\frac{\partial(\rho \vec{u})}{\partial t} + \text{div}\left(\rho\ \overrightarrow{u\vec{u}}\right) = \text{div}(\mu\ \text{grad}\,\vec{u}) + s_u$$

- **Energy Equation:**

$$\frac{\partial(\rho e)}{\partial t} + \mathrm{div}(\rho e\vec{u}) = \mathrm{div}(\mathrm{grad}\,T - P\vec{u}) + s_e$$

- **Species Transport Equation:**

$$\frac{\partial(\rho C_\alpha)}{\partial t} + \mathrm{div}(\rho\ C_\alpha\vec{u}) = \mathrm{div}(\rho D_\alpha\,\mathrm{grad}\,C_\alpha) + s_\alpha$$

where
ρ – density of fluid, kg/m^3;
P – pressure, Pa;
$\vec{u}$ – velocity vector of fluid element, m/s;
μ – viscosity of fluid, Pa·s;
e – total energy related to unit mass of fluid, kJ/kg;
λ – thermal conductivity, W/mK;
D_α – diffusion coefficient, m^2/s;
T – temperature, K;
C_α – concentration of species α in mixture, kmol/m^3.

S_m, S_u, S_h and S_α are the source terms in the above equations, which are associated with mass, momentum, energy, and transport of species.

- Mass Source term is considered in the continuity equation.
- Momentum Source term is considered in the momentum equation.
- Energy Source term is considered in the energy equation.
- Species Source term is considered in the species transport equation.

2.4 TWO-DIMENSIONAL GEOMETRY

A 2D solution geometry was prepared using the ANSYS Workbench software, and the geometry included three different sections, namely inlet, outlet and the reactor. The dimensions of all the sections were predefined in the 'Dimension' section of the modelling part of the Workbench. These dimensions are in correspondence with the experimental study and have been specified in the figure attached to the paper which shows the ex-situ reactor for 2D CFD study (Figure 2.1).

2.4.1 Meshing

The next step after the successful formation of the 2D geometry on the ANSYS DesignModeler is Meshing. As discussed above in the various methods of numerical analysis, the calculations are made at the interconnected

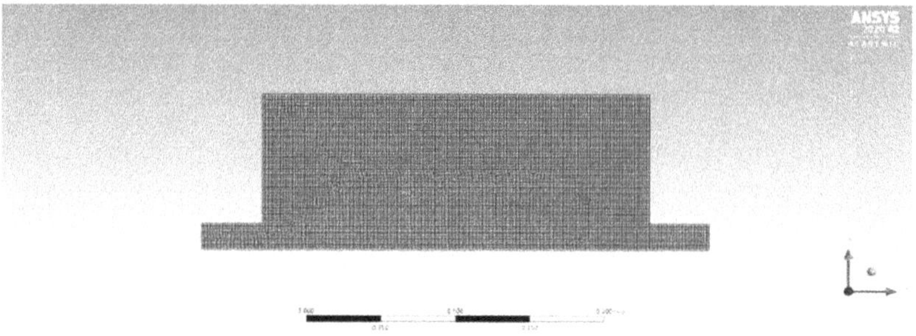

Figure 2.1 The 2D model with dimensions similar to the in situ reactor used in the experiment for the hydrogen-oriented underground coal gasification.

points known as nodes, and the overall geometry obtained along with these nodes is called a mesh. The number of nodes and elements in a specific geometry can be set which is visible in the Statistics section of the Details window. The number of nodes can be changed by changing the sizing values in accordance with the actual dimensions of the 2D geometry. The different sections of the geometry need to be assigned with a particular name to define the calculations. The named sections in our study are as follows:

1. Inlet
2. Walls
3. Base
4. Outlet

Generating Mesh is the next step, and this completes the first half of our study. The overall mesh with the specified number of nodes can be seen as attached to this chapter as a picture (Figure 2.2).

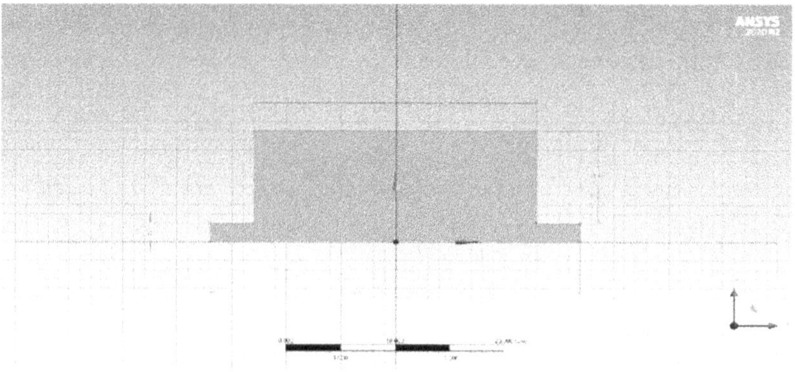

Figure 2.2 The mesh generated. The mesh has a total of 8024 nodes and 7808 elements which were further iterated for the grid independence analysis.

2.4.2 Solution model

One of the main steps in the numerical analysis of the 2D geometry representing Underground Coal Combustion is Initialisation & Calculations. These calculations in our study were done on ANSYS Fluent (Fluid Flow (FLUENT) Parallel Fluent) and the models used in the setup are as follows:

1. Energy model
2. Viscous model (k-epsilon)
3. Radiation model
4. Species Transport model
5. Discrete Phase model

The materials used in this study were the common combustion materials including coal and volatile air. The boundary conditions and the overall setup have been specified in the next section. The boundary layer conditions and the cell zone conditions have also been specified in Table 2.2.

The Species Transport model contains a coal calculator as well which gives the results for the Ultimate and Proximate analysis. The ultimate analysis takes place in accordance with the dry-ash-free basis (DAF). The coal properties have been tabulated in Table 2.3, and the overall results have been attached to the paper as the image (Figure 2.3).

Table 2.2 Boundary and initial conditions implemented in the numerical solution (Gür and Canbaz, 2020)

	Boundary conditions		*Initial conditions*
Inlet	Ignition	1000 K, 10 m^3/h, Air	Temperature 400 K
Oxygen	Oxygen-gasification	300 K, 3 m^3/h, Pure	Initial porosity 0.04 (% 4)
	Steam-gasification	400 K, 5 m^3/h, Steam	Initial permeability (1 mD)
			Coal moisture scalar 0.10011
Outlet	1 atm (absolute pressure)		Coal volatile scalar 0.347
			Coal fixed carbon scalar 0.3801

Table 2.3 Coal properties

Proximate analysis		*Ultimate analysis (DAF)*	
Volatile	0.5	Carbon	0.85
Fixed carbon	0.3	Hydrogen	0.1
Ash	0.1	Oxygen	0.04
Moisture	0.1	Nitrogen	0.01

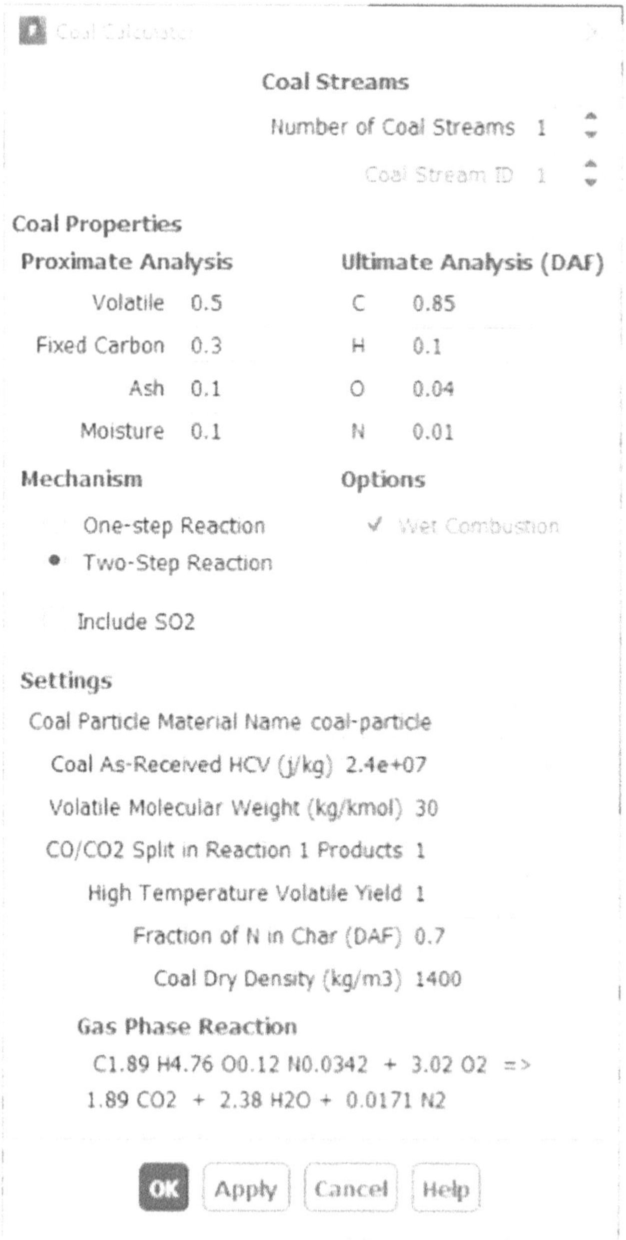

Figure 2.3 The coal calculator. The coal properties during the simulation study are mentioned in this which include the volatile content, fixed carbon, ash and moisture content during the proximate and ultimate analysis along with the mechanism of the reaction.

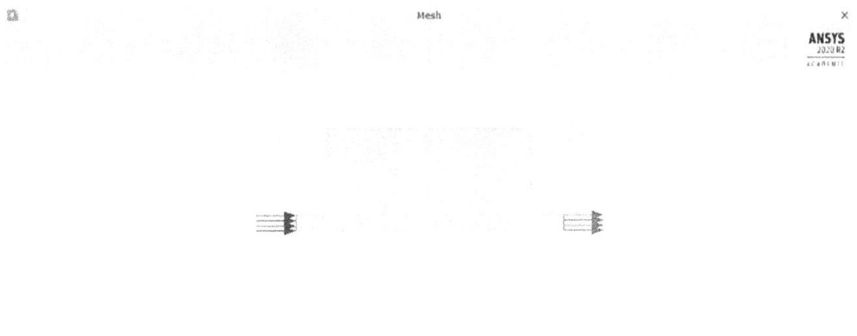

Figure 2.4 The overall model setup as obtained in FLUENT for the numerical analysis. The coal calculator is also found in the FLUENT module where the species and models are set for the calculations.

2.4.3 Overall model setup (Figure 2.4)

The process of meshing involves a number of conditions to be specified during the process. These include the boundary and initial conditions which are used for the numerical analysis during meshing. The main aim behind this is to specify the Flow Rates of the gasification agents in the process of UCG experiment that are implemented in the mathematical modelling. The total number of iterations performed during the simulation was 300, and the final graph and contour figure can be obtained as the images attached to the paper.

2.4.4 Grid sensitivity analysis

Any CFD results obtained from a simulation can never be trusted unless it is tested for dependence on the grid. As discussed in the above section of meshing, the relevance centre needs to be assigned between coarse, medium and fine. And the results obtained for a coarser mesh and finer mesh can never be the same. Therefore, the mesh needs to be varied so that an acceptable level of tolerance can be obtained and hence the **Grid Independence Test** comes into picture. The terms grid sensitivity analysis and grid independence test are often interchangeable.

This can be done by varying a set parameter related to the mesh size along with some output parameters. In our study, the 'Element Size – Number of Divisions' was compared with the output surface velocity and the output pressure. The data was plotted in a Comma Separated Values (CSV) (Figure 2.5) file and the point where the variation becomes negligible was taken as the optimum number of divisions.

2.5 SIMULATION ALGORITHM

To give the reader a more holistic view of the simulation study, an algorithm flowchart is attached to the paper which gives step-by-step information about the study.

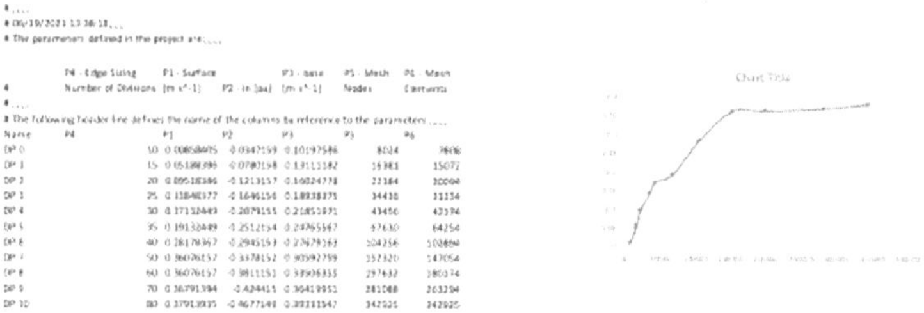
P4 - Edge Sizing
Number of Divisions
P1 - Surface
P5 - Mesh
Nodes
Name
P4
P1
P2
P3
P5
P6

Figure 2.5 The grid sensitivity analysis or the grid independence test.

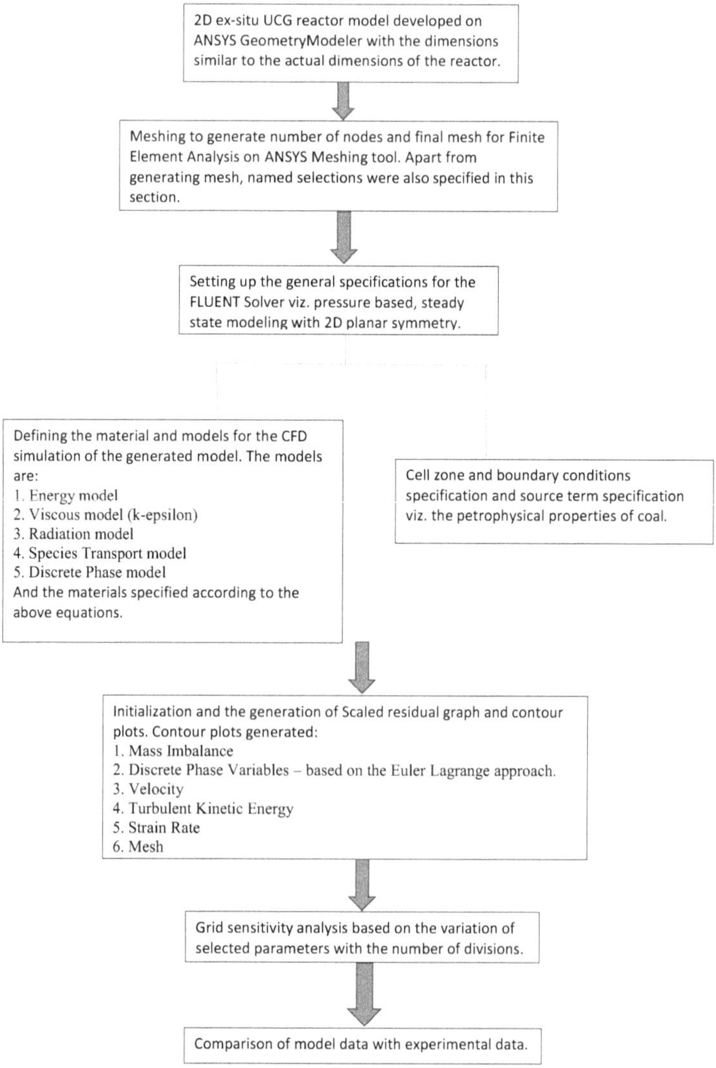
2D ex-situ UCG reactor model developed on ANSYS GeometryModeler with the dimensions similar to the actual dimensions of the reactor.
Meshing to generate number of nodes and final mesh for Finite Element Analysis on ANSYS Meshing tool. Apart from generating mesh, named selections were also specified in this section.
Setting up the general specifications for the FLUENT Solver viz. pressure based, steady state modeling with 2D planar symmetry.
Defining the material and models for the CFD simulation of the generated model. The models are:
1. Energy model
2. Viscous model (k-epsilon)
3. Radiation model
4. Species Transport model
5. Discrete Phase model
And the materials specified according to the above equations.
Cell zone and boundary conditions specification and source term specification viz. the petrophysical properties of coal.
Initialization and the generation of Scaled residual graph and contour plots. Contour plots generated:
1. Mass Imbalance
2. Discrete Phase Variables – based on the Euler Lagrange approach.
3. Velocity
4. Turbulent Kinetic Energy
5. Strain Rate
6. Mesh
Grid sensitivity analysis based on the variation of selected parameters with the number of divisions.
Comparison of model data with experimental data.

2.6 RESULTS

The graph shows the simulation of the UCG 2D model for a set of 300 iterations (Figure 2.6). The number of iterations can be varied as and when stability in the results is visible, which can be seen in this graph. The main approach behind this study is to satisfy a set of Partial Differential Equations which are mainly the following:

1. Energy equation
2. Momentum equation
3. Continuity equation
4. Species Transport equation.

All the above equations are based on and find applications in CFD.

Apart from the **Scaled Residual Graph** that shows the variation of all the residuals with the iterations, a set of **Contour Graphics** can also be obtained with the simulation.

The observed contour plots are as follows:

1. Mass Imbalance
2. Discrete Phase Variables – based on the Euler Lagrange approach
3. Velocity
4. Turbulent Kinetic Energy
5. Strain Rate
6. Mesh

The graphics of all these contours are attached to the paper (Figures 2.7–2.12). As shown by the contour figures, the experiment was initiated by supplying oxygen from the inlet to start and support the combustion. The supply of oxygen was characterised by parameters such as the flow rate, temperature at the inlet and pressure. All the parameters have been predefined in Table 2.2. The flow rate was a varied value of 3–5 m^3/h.

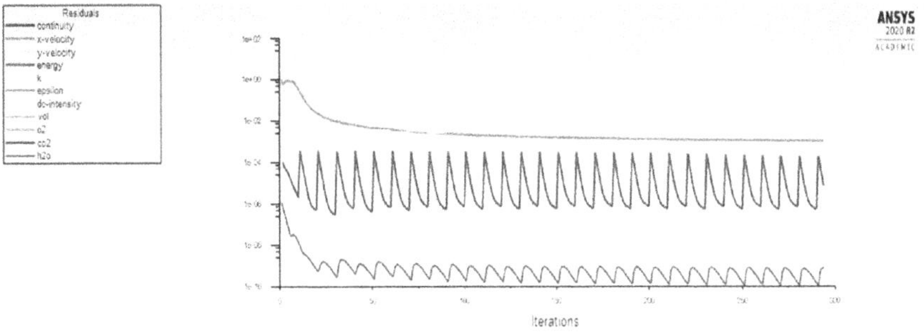

Figure 2.6 The plot for the overall simulation of the UCG model for a total of 300 iterations.

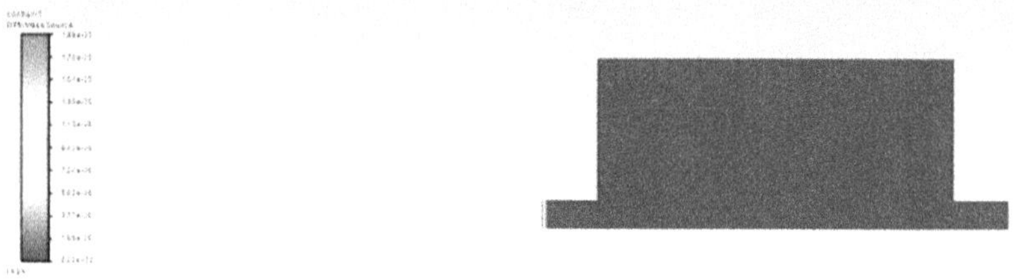

Figure 2.7 The contour diagram for the velocity variation over the reactor.

Figure 2.8 The contour diagram for the strain rate variation.

Figure 2.9 The contour diagram for the mesh.

Figure 2.10 The mass imbalance contour diagram representing the mass flow values at the interface.

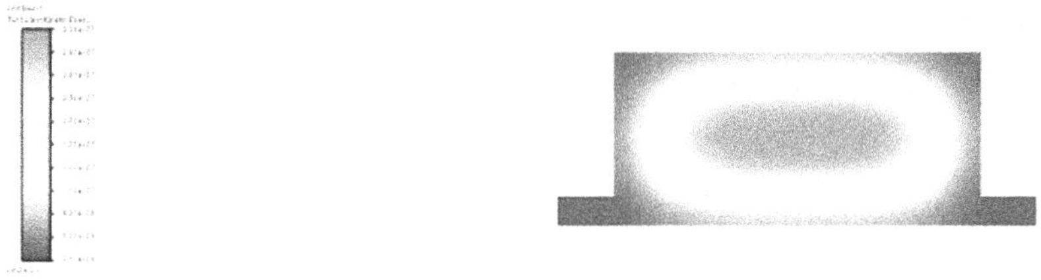

Figure 2.11 The turbulent kinetic energy contour diagram for the 2D UCG model.

Figure 2.12 The contour diagram for the discrete phase variables that are used for the continuous or the discrete phase flow.

The temperature was fixed to 400 k. The main aim behind performing these CFD simulations was to define the trends in CO_2 and H_2O which totally shows the dependence of water–gas shift (reaction 2.5).

2.7 CONCLUSION

Apart from the theoretical knowledge about the UCG process, this study also shed light on the numerical analysis of an experimental model and emphasised the technical aspects. The basic parameters such as the calorific values or the velocities of different species and their other properties used in the hydrogen-oriented UCG experiment and properties of the sample were brought in use for the 2D simulation of a UCG reactor. Initial parameters such temperature were set to 400 K, porosity 0.04 (4%), permeability (1 mD), coal moisture, volatile and fixed carbon content as 0.10011%, 0.347% and 0.3801%, respectively. The Fluent simulation gave various contour plots, namely Mass Imbalance, Discrete Phase Variables, Velocity, Turbulent Kinetic Energy, Strain Rate and Mesh. The contour diagrams show the variation of the above-mentioned parameters or the species used in the simulation over the reactor. These results were then compared with the previously conducted hydrogen-oriented UCG experiment, and there was not much variation in the results, and it presented a good agreement. According to the Scaled Residual Graph, the values of various parameters for the successive iterations showed a resemblance with the experimentally obtained values, which was the primary objective of our study.

REFERENCES

Anderson, J. (1995) Computational Fluid Dynamics: The Basics with Applications. McGraw Hill Inc.

Bhutto, A. W., Bazmi, A. A. and Zahedi, G. (2013) 'Underground coal gasification: From fundamentals to applications', Progress *in Energy and Combustion Science*, 39, 189–214. doi:10.1016/j.pecs.2012.09.004.

Coffield, D. and Shepherd, D. (1987) 'ANSY S FLUENT user's guide ANSYS', *Computer Communications*, 10(1), 21–29.

Gür, M. and Canbaz, E. D. (2020) 'Analysis of syngas production and reaction zones in hydrogen oriented underground coal gasification', *Fuel*, 269, 117331. doi:10.1016/j.fuel.2020.117331.

Laciak, M. et al. (2016) 'The analysis of the underground coal gasification in experimental equipment', *Energy*, 114, 332–343. doi:10.1016/j.energy.2016.08.004.

Li, Y., Liang, X. and Liang, J. (2007) 'An overview of the Chinese UCG program', *Data Science Journal*, 6(SUPPL.), S460–S466. doi:10.2481/dsj.6.S460.

Magnani, C. F. and Farouq Ali, S. M. (1975) 'Mathematical modeling of the stream method of underground coal gasification', *Society of Petroleum Engineers AIME Journal*, 15(5), 425–436. doi:10.2118/4996-pa.

Mocek, P. et al. (2016) 'Pilot-scale underground coal gasification (UCG) experiment in an operating mine "Wieczorek" in Poland', *Energy*, 111, 313–321. doi:10.1016/j.energy.2016.05.087.

Nourozieh, H. et al. (2010) 'Simulation study of underground coal gasification in Alberta reservoirs: Geological structure and process modeling', *Energy and Fuels*, 24, 3540–3550. doi:10.1021/ef9013828.

Perkins, G. (2018) 'Underground coal gasification - Part I: Field demonstrations and process performance', *Progress in Energy and Combustion Science*, 67, 158–187. doi:10.1016/j.pecs.2018.02.004.

Perkins, G. and Sahajwalla, V. (2007) 'Modelling of heat and mass transport phenomena and chemical reaction in underground coal gasification', *Chemical Engineering Research and Design*, 85(3A), 329–343. doi:10.1205/cherd06022.

Perkins, G. et al. (2016) 'Overview of underground coal gasification operations at Chinchilla, Australia', *Energy Sources, Part A: Recovery, Utilization and Environmental Effects*, 38(24), 3639–3646. doi:10.1080/15567036.2016.1188184.

Sarraf Shirazi, A., Karimipour, S. and Gupta, R. (2013) 'Numerical simulation and evaluation of cavity growth in in situ coal gasification', *Industrial and Engineering Chemistry Research*, 52(33), 11712–11722. doi:10.1021/ie302866c.

Yang, L. et al. (2008) 'Field test of large-scale hydrogen manufacturing from underground coal gasification (UCG)', *International Journal of Hydrogen Energy*, 33(4), 1275–1285. doi:10.1016/j.ijhydene.2007.12.055.

Żogała, A. and Janoszek, T. (2015) 'CFD simulations of influence of steam in gasification agent on parameters of UCG process', *Journal of Sustainable Mining*, 14(1), 2–11. doi:10.1016/j.jsm.2015.08.002.

Chapter 3

Modeling of MXene (Ti_3C_2) emerged blood flow through cosine shape stenosis

Niraj Rathore and N. Sandeep
Central University of Karnataka

3.1 INTRODUCTION

Nanofluids are a mixture of 1–100nm nanoparticles and base fluids. Combining the nanoparticles into the base fluids can improve the heat transfer rate proven by studies. Choi et al. [1] first demonstrated that nanoparticles can enhance the heat transfer rate of liquids. Nanoparticles are used in fields such as medical sciences, food industries, heating and cooling applications, etc. Thermal characteristics of nanofluids depend on base fluids and nanoparticles. Literature shows that nanoparticles display altered behavior with different base fluids. A case study of Magnetohydrodynamics (MHD) blood flow is discussed by Aasma et al. [2]. They noticed that the heat transfer rate and temperature of the blood flow could be improved by using carbon nanotubes, and volume fraction also plays a role. Copper nanoparticles emerged blood flowing through a catheterized mild stenosis artery with a thrombosis, analyzed by Thanna et al. [3] and see that increasing the volume fraction of nanoparticles can improve the blood flow. Shape, size, volume fraction, and other mechanical properties of the nanoparticles are also crucial in velocity and thermal results proved by researchers [4–10].

The optical and magnetic properties of nanoparticles can enhance the heat capacity of the tumor. The efficiency of cancer treatment in combination with thermal therapy has improved because of nanoparticles. The treatment of several diseases by nanoparticles using thermal therapy, drug delivery, bio-imaging, photo ablation therapy, biosensor, cancer treatment, and so on is possible. Magnetic nanoparticles such as iron oxides (Fe_3O_4), manganese (Mn), cobalt (Co), and nickel (Ni) are used for biomedical proposes due to their magnetic belongings such as high chemical constancy, nontoxicity, biocompatibility, and extraordinary susceptibility [11]. Metallic nanoparticles such as gold have optical properties, and a negative charge on the surface is helpful in surface functionalization. Patra et al. [12] fabricated gold nanoparticles for targeted therapy in cancer because of their large surface area to mass ratio and nontoxicity and biocompatibility. They

 DOI: 10.1201/9781032712079-3

discussed briefly the role of gold fine particles in targeted drug transport and thermal treatment of pancreatic cancer and found enormous potential to improve cancer treatment.

Similarly, bimetallic or alloy nanoparticles of iron–cobalt (Fe–Co) have high curie temperatures and high saturation magnetization. Still, biocompatibility is always a demand to overcome this problem—Fe–Co is coated with a biocompatible substance [13]. The metallic oxide nanoparticles, namely titanium oxide (TiO_2), are widely used for photocatalytic and photovoltaic devices. TiO_2 nanoparticles have numerous distinctive things, such as biocompatibility, chemical stability, and optical belongings [14]. Other metallic nanoparticles, such as MXene (Ti_3C_2) and cerium oxide (CeO_2), are also used for medical purposes. MXene is a two-dimensional (2D) molecule sheet obtained from carbides and nitrides of metals such as TiO_2 and was first synthesized in 2011 by Michel Naguib and Barsoum. MXene nanoparticles have higher electric conductivity and good physiochemical properties.

In the old days, arterial diseases could be treated with different methods, such as angioplasty or surgery. But nowadays, arterial disease treatment is possible without any surgery or cutting; this became possible because of nanoparticles. Nanoparticles and their thermophysical characteristics play a significant role in treatment. Moreover, the different shapes and sizes of the stenotic region display different velocities and thermal behavior. For this purpose, the study on different types of stenosis arteries and the impact of hybrid nanoparticles are made by Ashfaq et al. [14]. The different shapes, such as triangle, trapezoidal, bell shape, elliptical shape, composite, and irregular stenosis, are considered, and hybrid (Cu/Al_2O_3) blood flow is analyzed through it. They noticed that the flow nature and velocity of the fluids depend on the shape of the stenosis region; these findings are helpful in therapeutic strategy against arterial diseases. Researchers analyze different types and shapes of the stenotic region and suggest treatment through nanoparticles. Volumetric flow rate and heat transfer rate, and entropy in curved-shape artery is carried out by researchers and developed mathematical models to improve drug delivery method [15,16]. MXene nanoparticles have several favorable medical uses, such as in drug carriage, photothermic therapy, sensor, and cancer theranostics. Researchers conduct a detailed study of biomedical applications, synthesis, and recent MXene advancements [17,18]. Further biomedical, mechanical, and physical properties of MXene nanoparticles are discussed in the next section. This study contributes to the momentum and thermal behavior of MXene nanoparticles that emerged in blood flow in the stenosis artery. Other external impacts are applied and plotted graphically to analyze the thermal and velocity nature, which is helpful in regulating the flow and thermal behavior according to biomedical needs.

3.2 FEATURES AND PROPERTIES OF MXENE NANOPARTICLES

MXene composites have attractive properties such as big surface area, constancy, biocompatibility, high electric conductivity, energy harvesting, magnetism, etc., making them widely used 2-D materials. The electric conductivity of the Ti_3C_2 nanosheet is noticed (850 S/cm) without calcination, but it can be enhanced by increasing the temperature. After calcination at a temperature of 400°C, the electric conductivity can be expanded up to 70%. Moreover, by growing the calcination heat to 600°C, the conductivity is enhanced to 2140 S/cm [19]. A 2D shape MXene (Ti_3C_2) nanoparticles of lateral size 1–10 µm with 1 nm thickness has a density of 3700 kg/m^3 and heat capacity (C_p) of 2800 J/kg K [20]. Moreover, Ti_3C_2 has a vast surface extent that offers anchoring sites and allows an active build-up of toxins in tumor cells during cancer treatment. Advanced properties of MXene (Ti_3C_2) is noticed as compared with other 2-D material.

3.3 BIOMEDICAL APPLICATIONS OF MXENE NANOPARTICLES

MXene nanoparticles have excellent photothermal conversion efficiency, which makes them useful in cancer theranostics, photothermic elimination of cancer cells and ablation of cancer tumors, and tumor growth suppression. Photothermal agents can eliminate cancer cells, which convert light energy to heat energy. Ti_3C_2 nanoparticles are essential in MRI/CT imaging-guided thermal therapy [21]. Biomedical applications of MXene (Ti_3C_2) nanoparticles can be seen in Figure 3.1.

3.4 MATHEMATICAL MODELING

Consider the MXene (Ti_3C_2) emerging bloodstream in a curved stretching artery. The lower wall of the artery is considered radius δ, and the geometry shape depends on the δ values [15]. The cosine-shape stenosis is present in the artery, and the formulation is given below. The flow of blood-based nanofluids is taken along the s-direction with velocity u, and the r-axis is the upright direction of the stream displayed in Figure 3.2. The governing equations of momentum and temperature together with convective boundary conditions are developed to analyze velocity and thermal behavior due to external and internal effects.

The radial distance of cosine shape stenosis region from the center is given by [22]

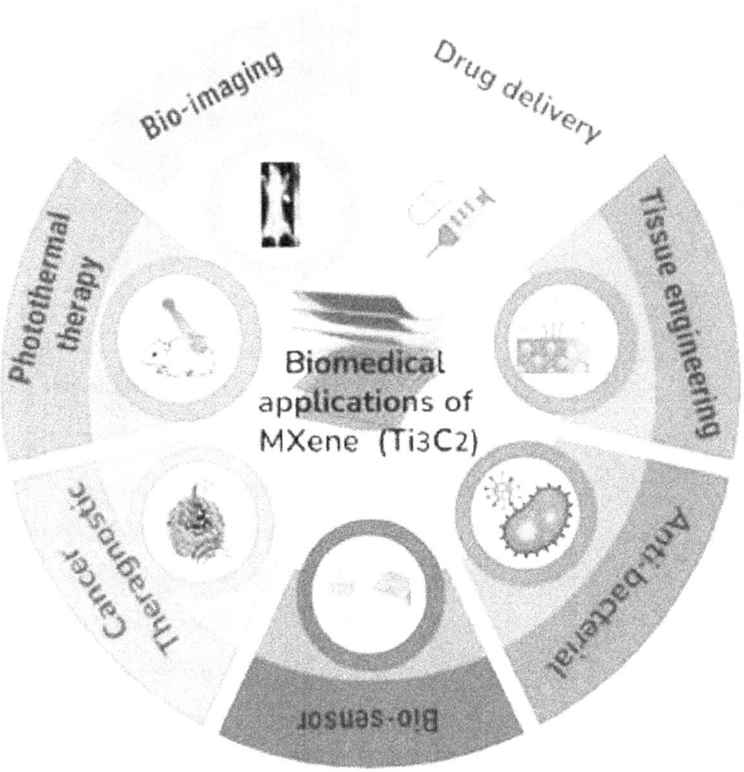

Figure 3.1 Biomedical applications of MXene (Ti_3C_2) nanoparticles.

$$R(s) = \begin{cases} R_0 - \dfrac{\lambda}{2}\left(1+\cos\left(\dfrac{4\pi s}{L_0}\right)\right), & -\dfrac{L_0}{4} < s < \dfrac{L_0}{4} \\ R_0, & \text{otherwise} \end{cases}.$$

Here it is assumed that the length of the stenosis is $\frac{L_0}{2}$ in s-direction.

3.4.1 Flow analysis

3.4.1.1 Equation of continuity

$$\frac{\partial}{\partial r}\left((\delta + r)w\right) + \delta\frac{\partial u}{\partial s} = 0. \tag{3.1}$$

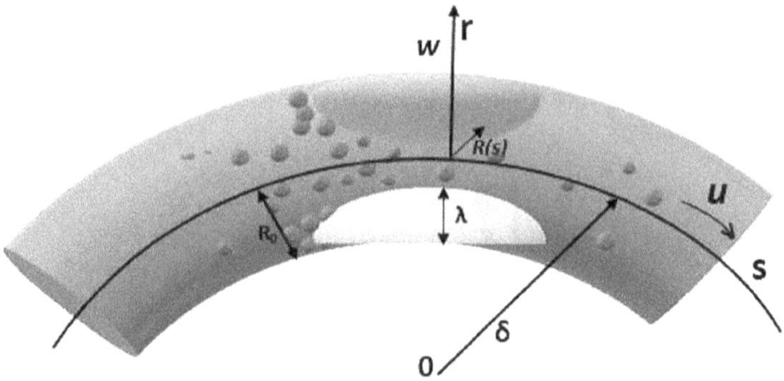

Figure 3.2 Schematic diagram of a curved artery of R_0 is the radius of the artery (without stenosis), λ is the maximum height of the stenosis region, and $R(s)$ is the variable radial distance of the stenotic region from the s-axis.

3.4.1.2 Pressure term

$$\frac{1}{\rho_{nf}}\frac{\partial p}{\partial r}-\frac{1}{r+\delta}u^2=0, \tag{3.2}$$

where u, w are velocity components in s, r ways, respectively.

3.4.1.3 The momentum of the blood flow

$$\left(\begin{array}{l} w\dfrac{\partial u}{\partial r}+\dfrac{\delta u}{\delta+r}\dfrac{\partial u}{\partial s} \\ +\dfrac{uw}{\delta+r}\end{array}\right)+\frac{\delta}{\delta+r}\frac{\partial p}{\partial s}=\left(1+\frac{1}{\gamma}\right)\upsilon_{nf}\left(\begin{array}{l}\dfrac{\partial^2 u}{\partial r^2}+\dfrac{1}{\delta+r}\dfrac{\partial u}{\partial r} \\ -\dfrac{u}{(\delta+r)^2}\end{array}\right)$$

$$-\frac{C^* u^2}{\rho_{nf}k_f\sqrt{k}_1}+\frac{g(\rho\beta)_{nf}}{\rho_{nf}}\left(T-T_{w_1}\right). \tag{3.3}$$

The momentum boundary conditions are considered

$$u=a\exp\left(\frac{s}{L}\right),\quad w=0,\quad \text{at}\quad r=0;\quad \text{and}\quad \frac{\partial u}{\partial r}\to 0,\quad u\to 0,\quad \text{as}\quad r\to R(s). \tag{3.4}$$

Taking similarity transformation to convert the above equations into a non-dimensional form

Table 3.1 Thermophysical ratios of the nanofluids [23]

Features	*Nanofluid parameter*
Density	$\frac{\rho_{nf}}{\rho_f}=(1-\phi)+\frac{\phi\rho_{sp}}{\rho_f}$
Viscosity	$\frac{\mu_{nf}}{\mu_f}=(1-\phi)^{-2.5}$
Heat capacity	$\frac{(\rho C_p)_{nf}}{(\rho C_p)_f}=(1-\phi)+\frac{\phi(\rho C_p)_{sp}}{(\rho C_p)_f}$
Thermal conductivity	$\frac{k_{nf}}{k_f}=\frac{k_{sp}+k_f(m-1)+(k_{sp}-k_f)(m-1)\phi}{k_{sp}+k_f(m-1)+\phi(k_f-k_{sp})}$
Thermal expansion	$\frac{(\rho\beta)_{nf}}{(\rho\beta)_f}=(1-\phi)+\phi\frac{(\rho\beta)_{sp}}{(\rho\beta)_f}$

$$\left.\begin{array}{l}\xi=r\left(\dfrac{a\exp\left(\dfrac{s}{L}\right)}{2\upsilon_f L}\right)^{1/2},\quad p=\rho_f a^2\exp\left(\dfrac{s}{L}\right)P(\xi),\quad u=a\exp\left(\dfrac{s}{L}\right)\dfrac{\partial f}{\partial\xi}\\ w=-\dfrac{\delta}{r+\delta}\left(\dfrac{a\upsilon_f}{2L}\right)^{1/2}\left(f(\xi)+\xi\dfrac{\partial f}{\partial\xi}\right)\exp\left(\dfrac{s}{L}\right)\end{array}\right\}.\quad(3.5)$$

Here, f denotes fluid, sp is solid, and nf is nanofluid.

By using equations (3.2), (3.5), and Table 3.1, equation (3.3) converts into the following form:

$$\left.\begin{array}{l}\left(\dfrac{\rho_{nf}}{\rho_f}\right)\left(\dfrac{\kappa}{(\xi+\kappa)^2}ff''+\dfrac{\kappa}{(\xi+\kappa)}ff'''-\dfrac{\kappa}{(\xi+\kappa)^3}ff'-\dfrac{3\kappa}{(\xi+\kappa)^2}f'^2-\dfrac{3\kappa}{(\xi+\kappa)}f'f''\right)\\ +\left(1+\dfrac{1}{\gamma}\right)\dfrac{\mu_{nf}}{\mu_f}\left(f''''+\dfrac{2}{(\xi+\kappa)}f'''-\dfrac{1}{(\xi+\kappa)^2}f''+\dfrac{1}{(\xi+\kappa)^3}f'\right)-Gr\dfrac{(\rho\beta)_{nf}}{(\rho\beta)_f}\theta+Frf'^2=0\end{array}\right\}.\quad(3.6)$$

Moreover, dimensional boundary conditions in equation (3.4) transform into the non-dimensional form

$$\frac{\partial f}{\partial\xi}=1,\quad f=0,\quad \text{at}\quad \xi=0,\quad \text{and}\quad \frac{\partial f}{\partial\xi}\to 0,\quad \frac{\partial^2 f}{\partial\xi^2}\to 0,\quad \text{as}\quad \xi\to R_{w1},\quad(3.7)$$

Table 3.2 Thermal belongings of base liquid and nanoparticle [19–20,26]

	ρ(kg/m)	C_p (J/kgK)	K(W/mK)	β(1/K)
Ti_3C_2	3700	2800	55.8	8.9×10^{-6}
Blood	1050	3617	0.52	0.8×10^{5}

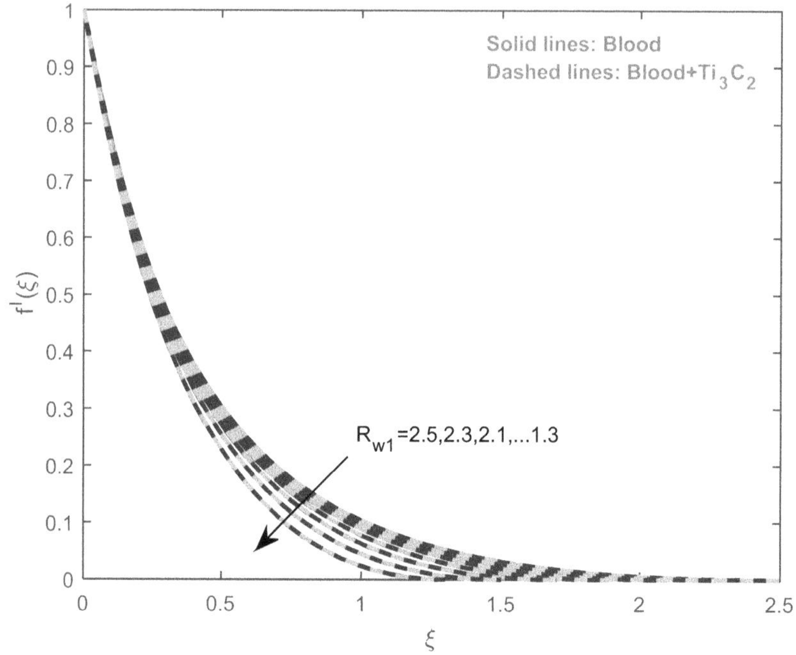

Figure 3.3 Velocity of fluids for decreasing radial distance values from the artery's origin.

where $R_{w_1} = R(s)\left(\dfrac{a\exp\left(\dfrac{s}{L}\right)}{2\upsilon_f L}\right)^{1/2}$ is variable radial distance from the center of the artery, $Fr = \dfrac{C^* s}{\rho_f \sqrt{k_1}}$ is Forchheimer number [24].

3.4.2 Thermal analysis

$$\left.\left(\frac{\delta u}{\delta + r}\frac{\partial T}{\partial s} + w\frac{\partial T}{\partial r}\right) = \alpha_{nf}\left(\frac{1}{\delta + r}\frac{\partial T}{\partial r} + \frac{\partial^2 T}{\partial r^2}\right) - \frac{1}{(r+\delta)(\rho c_p)_{nf}}\frac{\partial}{\partial r}(\delta + r)q_1 + \frac{q'''}{(\rho c_p)_{nf}}\right\}. \tag{3.8}$$

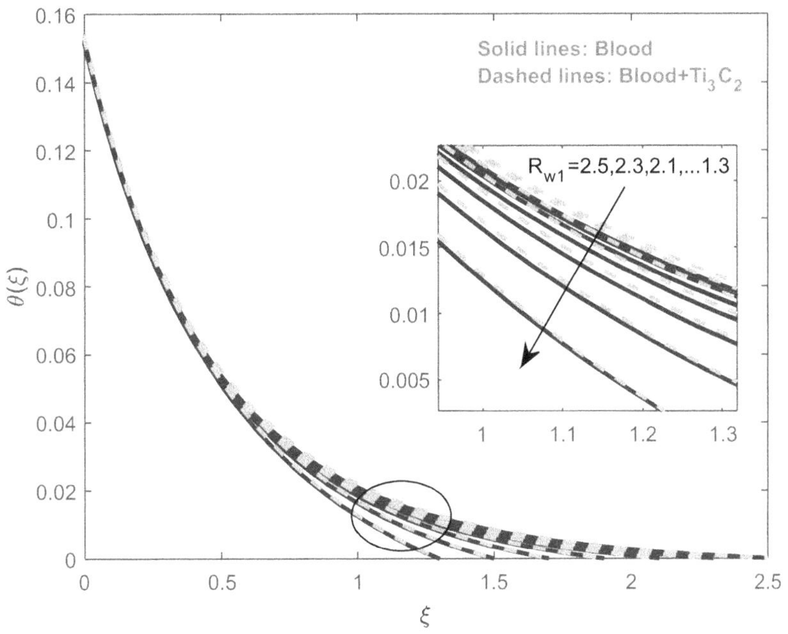

Figure 3.4 Temperature of fluids for decreasing radial distance values from the artery's origin.

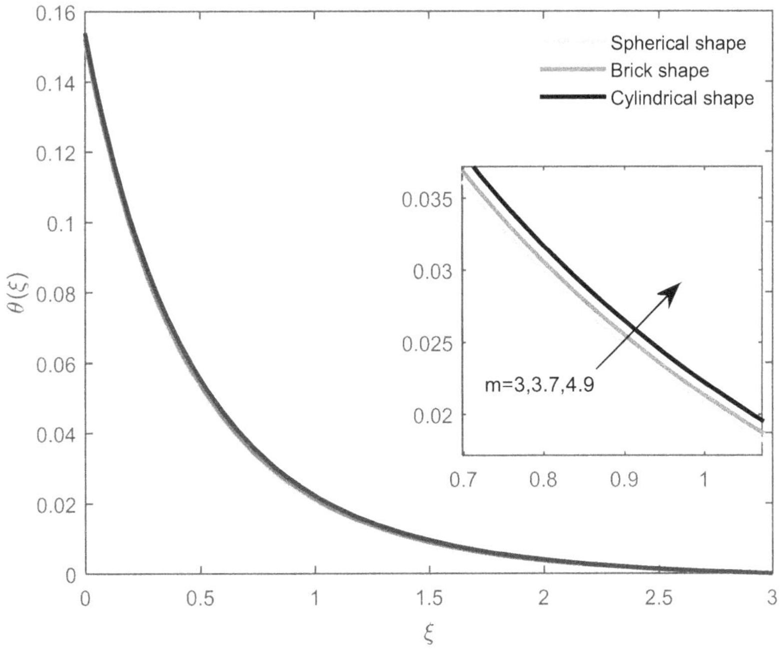

Figure 3.5 Thermal difference for different shapes of MXene (Ti_3C_2) nanoparticles.

With the BCs

$$k_{nf}\frac{\partial T}{\partial r}=h_f(T-T_0),\quad \text{at}\quad r=0;\quad \text{and}\quad T\rightarrow T_{w_1},\quad \text{as}\quad r\rightarrow R(s). \tag{3.9}$$

The temperature similarity variable is considered as

$$T=\theta(\xi)\left(T_0-T_{w_1}\right)+T_{w_1}. \tag{3.10}$$

Here, $q_1=-\dfrac{4\sigma^*}{3k^*}\dfrac{\partial T^4}{\partial r}$ is Rosseland radiation heat flux and q''' is an uneven heat sink or source [25] and given as $q'''=\dfrac{k_{nf}u_w(T_0-T_{w_1})}{2(\rho c_p)_{nf}L\upsilon_f}\left(A*f'+B*\dfrac{(T-T_{w_1})}{(T_0-T_{w_1})}\right)$.

Using equation (3.10) and Table 3.1, equations (3.8) and (3.9) transformed in non-dimensional form

$$\left(\frac{k_{nf}}{k_f}+Ra\right)\left(\theta''+\frac{1}{\xi+\kappa}\theta'\right)+Pr\frac{(\rho c_p)_{nf}}{(\rho c_p)_f}\frac{\kappa}{\xi+\kappa}(f\theta'-f'\theta)+(A*f'+B*\theta)=0. \tag{3.11}$$

The converted limits are

$$\frac{\partial\theta}{\partial\xi}=-Bi(1-\theta),\quad \text{at}\quad \xi=0,\quad \text{and}\quad \theta\rightarrow 0,\quad \text{as}\quad \xi\rightarrow R_{w_1}. \tag{3.12}$$

Friction at the arterial wall is computed by [26]

$$\mathrm{Re}_s^{1/2}C_{fs}=\frac{\mu_{nf}}{\mu_f}\left(f''(0)-\kappa^{-1}f'(0)\right). \tag{3.13}$$

Thermal transference of the nanofluids is computed by local Nusselt quantity and specified [27] by

$$\mathrm{Re}_s^{-1/2}Nu_s=-\left(\frac{k_{nf}}{k_f}+Ra\right)\theta'(0), \tag{3.14}$$

where the local Reynolds numeral is $\mathrm{Re}_s=\dfrac{aL\exp(s/L)}{\upsilon_f}$.

The velocity and thermal profiles of the blood flow inside the curved-shape artery can be examined with the coupled equations (3.6) and (3.11) together with boundary conditions (3.7) and (3.8). Moreover, the drag on the wall and heat transmission rate can be calculated with the assistance of equations (3.13) and (3.14), respectively. We developed the MATLAB code to solve the above dimensionless couples nonlinear ODE and plotted the results.

3.5 RESULTS AND DISCUSSION

A mathematical model of the 2D flow of nanofluids flowing through cosine shape stenosis arteries is developed to analyze thermal and flow behavior for different physical parameters. The heat transfer rate, temperature, and velocity of the base fluids and nanofluids are compared to observe the impact of MXene (Ti_3C_2) nanoparticles. Governing flow equations are converted into ordinary differential equations by appropriate similarity conversions and solved mathematically using the bvp5c MATLAB package. For the computation purpose and converging profiles of momentum and thermal boundaries, the physical parameters are taken in the range of $m = 4.9$, $\kappa = 3$, $Gr = 0.3$, $Ra = 1$, $A^* = 0.01$, $B^* = 0.01$, $Bi = 0.4$, $\gamma = 0.3$, $Fr = 0.3$. The outcomes of the study are shown graphically and in tabular form. Figures 3.3 and 3.4 exhibit the momentum and energy for different heights of stenosis formed in an artery. The thermal difference and heat transfer rate for MXene (Ti_3C_2) nanoparticle shapes are exhibited in Figures 3.5 and 3.6. The influence of different physical constraints on velocity and heat is displayed in Figures 3.7–3.15, and a comparison of heat transfer rate in pure blood and nanoparticles mixed blood is shown in Figures 3.16 and 3.17. Moreover, drag and Nusselt number values are depicted in Table 3.2. The detailed information is explained below.

3.5.1 Impact on velocity

Relation between a radial distance of the stenosis region from the center of the artery. For decreasing values of radial distance from the origin of the artery (i.e., increment the atherosclerosis height), velocity and thermal change are described in Figures 3.3 and 3.4. It is noticed that with decreasing

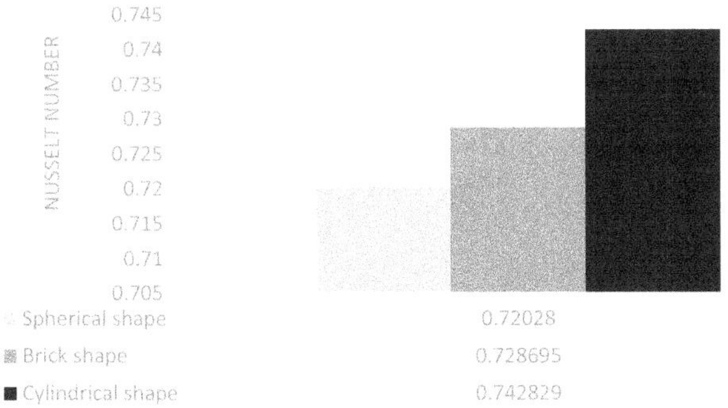

Figure 3.6 Heat transfer rate for different MXene (Ti_3C_2) particle shapes.

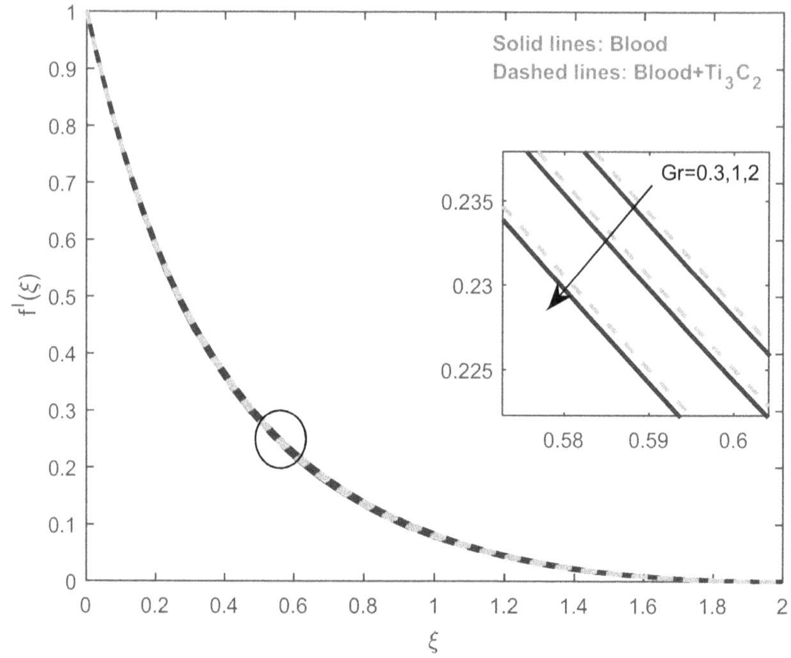

Figure 3.7 Velocity outlines for increasing values of Grashof number.

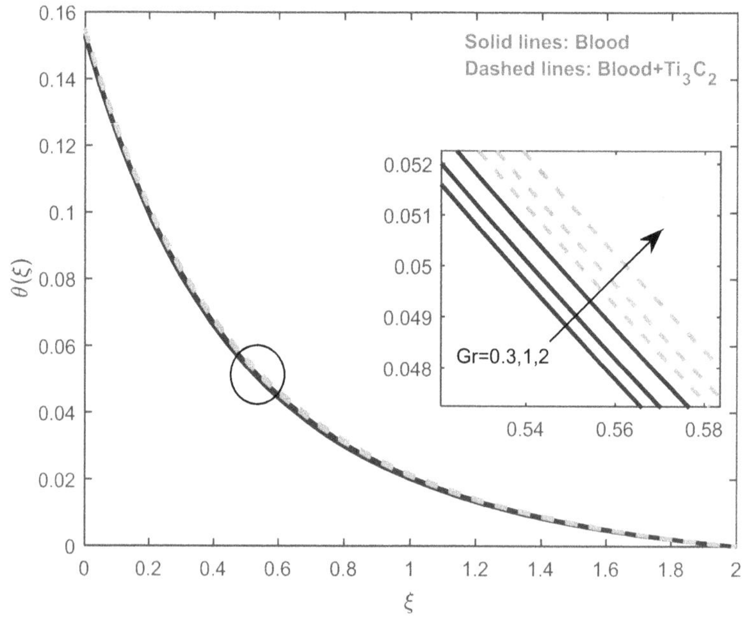

Figure 3.8 Change in thermal profiles for increasing values of Grashof number *Gr*.

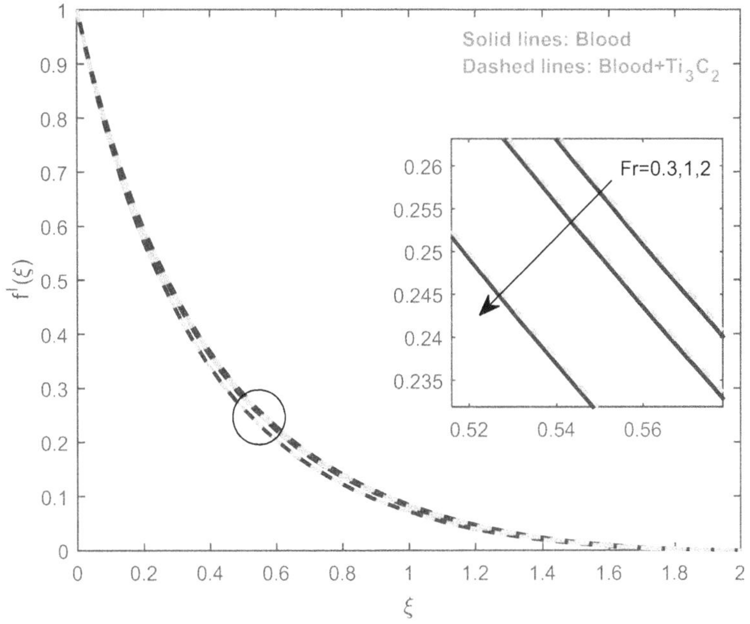

Figure 3.9 Velocity profiles for increasing values of Forchheimer number *Fr*.

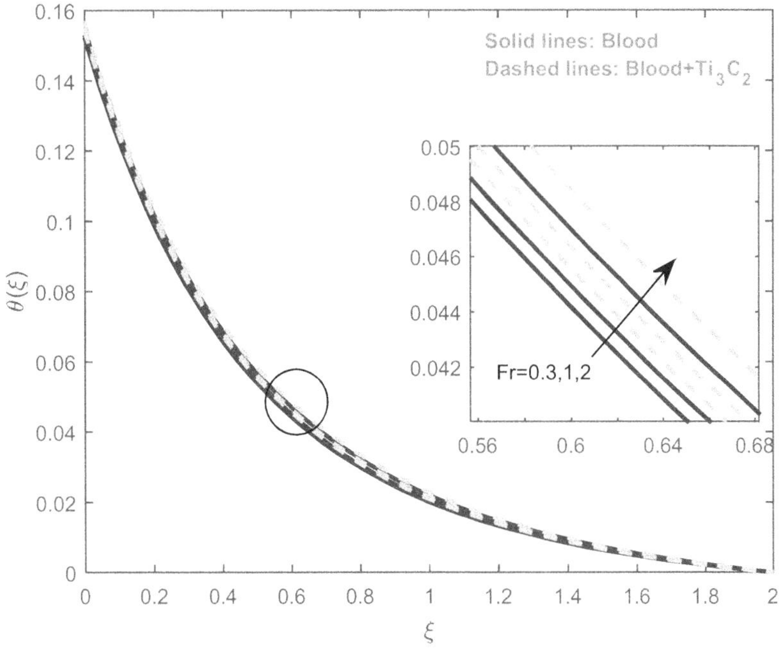

Figure 3.10 Thermal profiles for increasing values of Forchheimer number *Fr*.

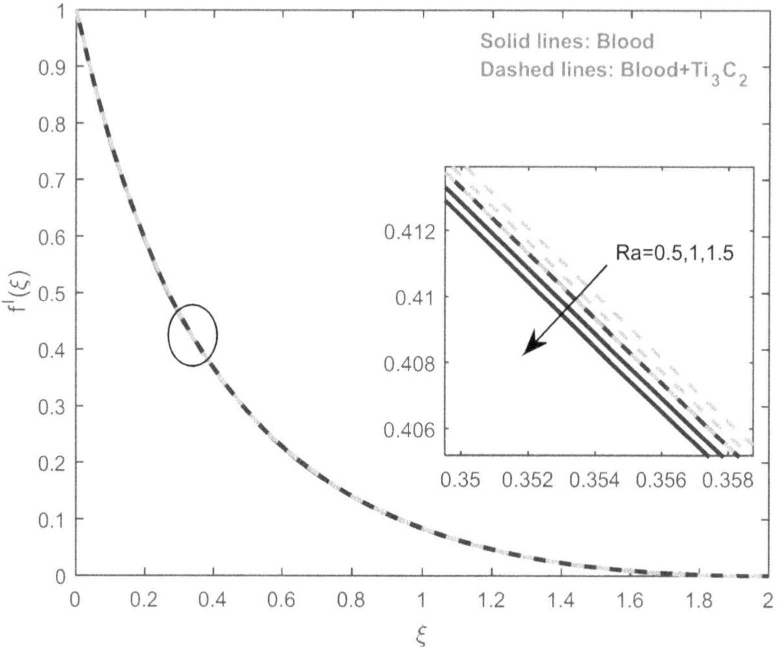

Figure 3.11 Velocity profiles for increasing values of thermal radiation parameters.

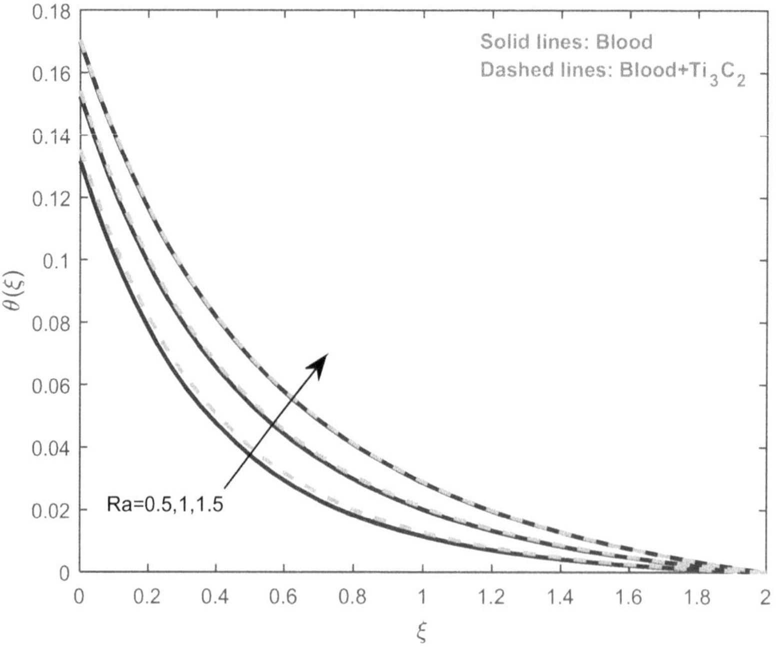

Figure 3.12 Temperature values for increasing values of thermal radiation parameters.

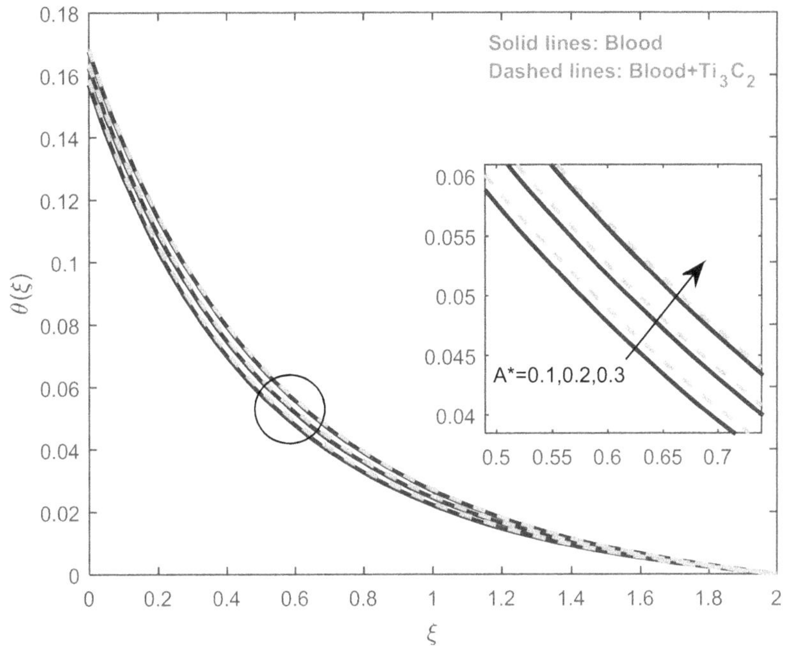

Figure 3.13 Thermal profiles for growing values of heat source/sink values A^*.

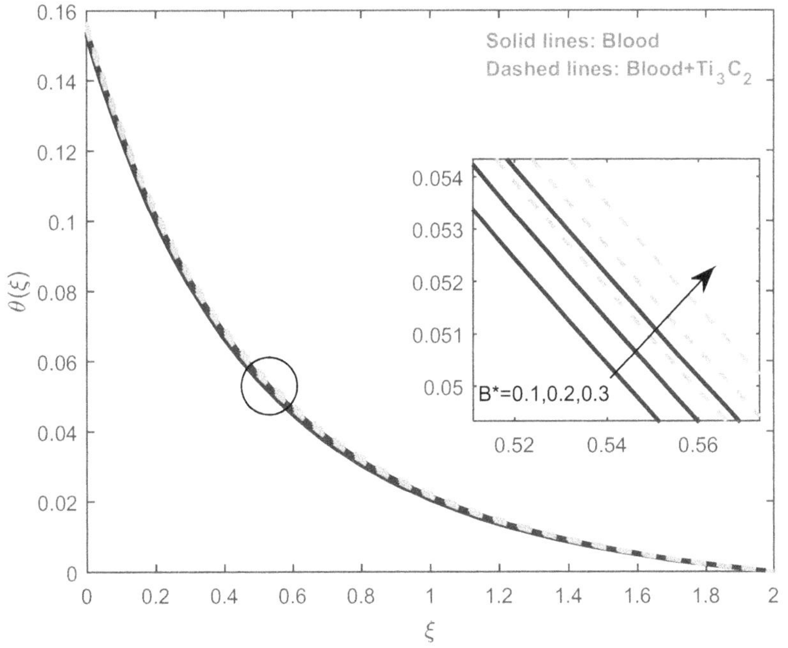

Figure 3.14 Thermal nature for increasing values of heat sink/source parameters B^*.

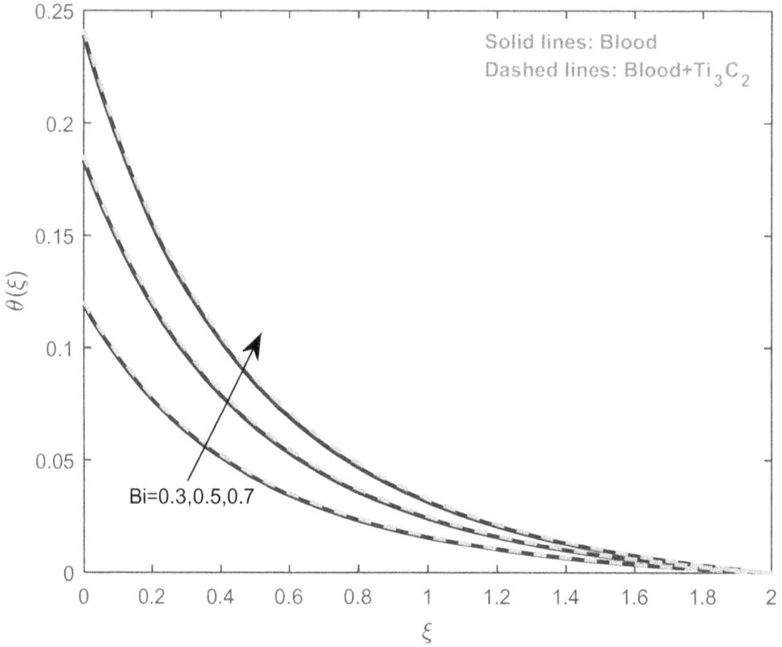

Figure 3.15 Thermal profiles for increasing values of Biot number *Bi*.

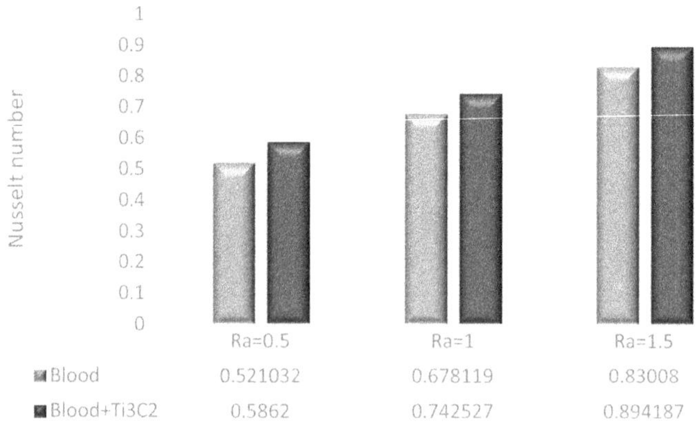

Figure 3.16 Comparison of heat transfer rate in blood and nanoblood flow for different thermal radiation parameters.

values of radial distance from the origin of the artery, the velocity and temperature are reduced. Hence, the increment in the stenosis heights of the artery affects the blood flow, which can cause heart attack and other arterial diseases. For growing numerals of Grashof (Gr) number, the moment of the flow declines, but the temperature improves displayed as shown in Figure 3.7. Moreover, it is noticed that the velocity is enhanced by mixing

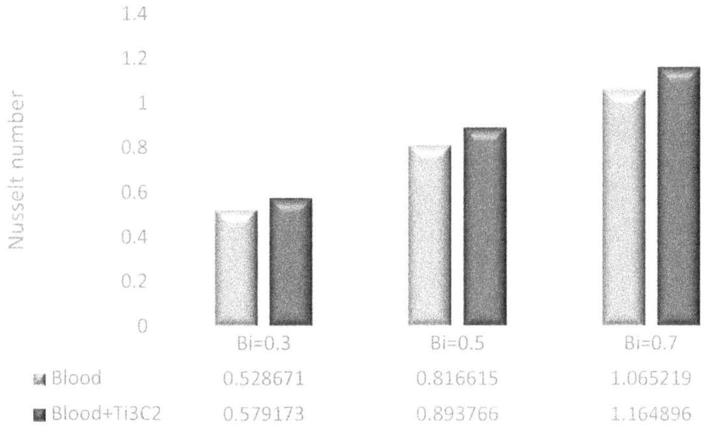

Figure 3.17 Comparison of heat transfer rate for blood and nanoblood flow for different Biot numbers.

MXene (Ti_3C_2) nanoparticles into the blood flow since the Grashof number is a non-dimensional parameter that characterizes the comparative importance of buoyancy forces to viscous forces in a liquid flow. Figure 3.9 indicates the velocity change for increasing values of the Forchheimer number. The velocity is getting slow for Forchheimer numbers. Moreover, MXene nanoparticles help to improve the momentum of the fluid. As the Darcy–Forchheimer number rises, it suggests that the additional inertial resistance due to the presence of the porous medium becomes more significant. This increased resistance slows the flow velocity and may be the reason for the above trend. When the Darcy–Forchheimer number rises, it indicates a higher contribution of inertial forces than viscous forces in the flow. The thermal radiation effect is applied over the blood flow to analyze flow and thermal changes; the results for the same are displayed in Figures 3.11 and 3.12. Decay in the velocity is noticed for growing values of thermal radiation parameters shown in Figure 3.11. The MXene nanoparticles do not change the velocity significantly of the blood flow under the impact of thermal radiation.

3.5.2 Impact on temperature

The relation between decreasing values of the radial distance of the stenosis region from the center of the artery (i.e., increment atherosclerosis) and thermal change is described in Figure 3.4. It is noticed that with decreasing values of radius at the stenosis region, the temperature is reduced. This may cause a blood clot in the artery. Figure 3.5 represents the temperature of fluids due to the different shape of MXene nanoparticles. The cylindrical shape nanoparticles have higher temperature and heat transfer rates when compared with spherical nanoparticles. This is the reason we used cylindrical shape nanoparticles to observe other results. For increasing values of Grashof (Gr) number, the temperature improves displayed in Figure 3.8. Moreover, velocity and temperature are improved by mixing MXene nanoparticles into the blood flow. Figure 3.10 indicates temperature change for increasing values of the Forchheimer number. The temperature is enhanced for Forchheimer numbers. Moreover, MXene nanoparticles help to improve the fluid's velocity and temperature. The thermal radiation effect is applied over the blood flow to analyze flow and thermal changes; the results for the same are displayed in Figures 3.11 and 3.12. Improvement in temperature is noticed for growing values of thermal radiation parameters, as shown in Figure 3.12. The MXene nanoparticles help improve the blood flow temperature for the thermal radiation effect. Thermal analysis of fluids due to heat source and sink parameters are shown in Figures 3.13 and 3.14. The flow temperature is increased for increasing heat source/sink parameter values, and the temperature of MXene nanoparticles mixed blood flow is higher than usual. Figure 3.15 displays progress in temperature for growing values of Biot numbers. Moreover, it is observed that the temperature of nanofluids is higher than of pure blood flow. If increasing the temperature leads to an upsurge in the Biot numeral, it implies that the importance of internal thermal resistance (conduction) is becoming relatively more significant compared with the exterior thermal resistance (convection).

3.5.3 Impact on the heat transfer rate

The heat transmission rate for different shape of nanoparticles are discussed in Figure 3.6; it is observed that the heat transmission rate is higher for cylindrical shape nanoparticles when compared with spherical shape nanoparticles. Therefore, in our study, we used cylindrical shape MXene (Ti_3C_2) nanoparticles. Thermal radiation is one of the mechanisms by which heat can be transferred between objects or surfaces at different temperatures. In fluid dynamics, the presence of thermal radiation can have significant effects on heat transfer rates. We explore how thermal radiation impacts heat transfer in fluids. We can see in Figure 3.16 the heat transfer rate due to growing values of thermal radiation factors. Heat transfer rate upsurges for cumulative standards of thermal radiation factors. Moreover, it is noticed that the heat transfer rate of MXene mixed blood flow is higher than the usual blood flow. Similar behavior is observed for Biot numbers, that is, the heat transfer rate increases for increasing values of Biot numbers.

Moreover, nanofluids have a higher heat transfer rate than pure blood flow. Biot number parameters are noticed in and displayed in Figure 3.17. It is observed that the heat transmission rate upsurges for thermal radiation parameters, and MXene-based blood flow has a higher heat transfer rate than usual.

Table 3.3 Drag coefficient and Nusselt numerals for different physical parameters

Gr	*Fr*	*Ra*	*A**	*B**	*Bi*	*Blood*	*Blood + Ti_3C_2*	*Blood*	*Blood + ≈Ti_3C_2*
0.3						−3.064624	−3.386033	0.678119	0.742527
1.0						−3.090157	−3.413727	0.677827	0.742207
2.0						−3.127055	−3.453738	0.677397	0.741735
	0.3					−3.064624	−3.386033	0.678119	0.742527
	1.0					−3.122668	−3.450169	0.677527	0.741869
	2.0					−3.273227	−3.616568	0.675960	0.740127
		0.5				−3.062301	−3.383769	0.521032	0.586200
		1.0				−3.064624	−3.386033	0.678119	0.742527
		1.5				−3.066628	−3.388005	0.830080	0.894187
			0.1			−3.065171	−3.386574	0.674326	0.738665
			0.2			−3.065778	−3.387173	0.670112	0.734375
			0.3			−3.066385	−3.387773	0.665900	0.730085
				0.1		−3.064724	−3.386134	0.677442	0.741824
				0.2		−3.064838	−3.386249	0.676673	0.741025
				0.3		−3.064955	−3.386367	0.675883	0.740206
					0.3	−3.062239	−3.383450	0.528671	0.579173
					0.5	−3.066836	−3.388428	0.816615	0.893766
					0.7	−3.070812	−3.392725	1.065219	1.164896

Table 3.3 represents the heat allocation rate and drag in nanofluids corresponding to different parameters. It is noticed that drag due to MXene (Ti_3C_2) nanoparticles is reduced, and the heat transfer rate is enhanced for all parameters.

3.6 CONCLUSIONS

Two-dimensional unsteady incompressible blood flow in cosine shape stenotic curved artery is observed. The mathematical formulations are constructed to analyze flow in curved-shape artery, and boundary conditions are developed for cosine shape atherosis. The external pressure and force are applied in the form of Darcy–Forchheimer and buoyance terms to analyze the flow behavior. The impact of thermal radiation and uneven heat generation and absorption parameters are considered in the energy equation to observe the thermal behavior of the flow. The convective boundary condition is assumed to analyze the relation between convective and conductive heat and the direction of the heat transfer. The MXene nanoparticles are taken because of their big surface region, admirable electrical conductivity, good biocompatibility, and tunable physicochemical belongings. The momentum, energy, drag, and heat allocation rate are compared for pure blood flow and MXene (Ti_3C_2) mixed blood flow under different parameters. The results are obtained numerically with the help of MATLAB software and exhibited through graphs and tables. The major observations are given below:

- The blood flow heat transfer rate is improved after mixing the MXene (Ti_3C_2) nanoparticles. The faster rate of heat transfer can be favorable in thermal therapy.
- The drag of the blood flow is reduced due to MXene nanoparticles. Therefore, the drug delivery method can be improved using the MXene nanoparticles.
- Nanofluids' temperature and heat transfer rate are enhanced by growing values of thermal radiation and Biot parameters. This phenomenon is helpful in thermal management to eradicate the tumor.
- The energy source/sink is also valuable for improving heat at a targeted position; this can be beneficial to abolish undesirable tissues using thermal ablation.
- MXene nanoparticles help to improve temperature and heat transfer rate without much changes in velocity; these results are essential in the thermal treatment of diseases or tumors without disturbing healthy tissues.
- MXene (Ti_3C_2) nanoparticles are advantageous in thermal ablation techniques. Moreover, external physical factors help to control momentum and heat. This occurrence can be used in the treatment of stenotic arterial diseases.

NOMENCLATURE

$\kappa = \delta\left(\dfrac{a\exp\left(\dfrac{s}{L}\right)}{2\upsilon_f L}\right)^{1/2}$	Curvature parameter of artery
γ	Casson fluid parameter
λ	Height of the stenosis
μ_{nf}	Nanoparticle's viscosity
$(\rho C_p)_{nf}$	Nanoparticle's heat capacity
ρ	Density
ρ_{nf}	Nanoparticle's density
σ	Electrical conductivity
ϕ_1	Volume fraction of MXene (Ti_3C_2) nanoparticles

υ_{nf}	Kinematic viscidness
A^*, B^*	Heat source/sink parameter
C_{fs}	Skin friction coefficient
C_p	Heat capacity
$\exp\left(\frac{s}{L}\right)$	Exponential velocity
K	Thermal conductivity
k_{nf}	Nanoparticle's thermal conductivity
$L_0 / 2$	Length of an artery
m	Nanoparticle's shape factor
nf	Nanofluid particle
Nu_s	Heat transmission factor
$\Pr = \frac{\upsilon_f}{\alpha_f}$	Prandtl number
q_1	Radiative heat transfer
R_0	Radius of healthy artery
$Ra = \frac{16\sigma^* T_\infty^3}{3k_f k^*}$	Thermic radiation
$R(s)$	Radial distance of stenosis from origin
s	Solid particle
T_0	Temperature at the center of artery
T_{w_1}	Temperature at athlerosis wall
$Bi = h_f \sqrt{\frac{2\upsilon_f L}{a\exp(\frac{s}{L})}}$	Biot number
$Fr = \frac{C^* s}{\rho_f \sqrt{k_1}}$	Forchheimer number
$Gr = \frac{g\beta_f \left(T_{w_1} - T_\infty\right) s^3}{\upsilon_f^2}$	Grashoff number

REFERENCES

[1] Choi, S. U.S., and Eastman, J. A. (1995). Enhancing thermal conductivity of fluids with nanoparticles". *ASME International Mechanical Engineering Congress & Exposition, November 12–17, 1995*, United States. https://www.osti.gov/servlets/purl/196525

[2] Khalid, A., Khan, I., Khan, A., Shafie, S., & Tlili, I. (2018). Case study of MHD blood flow in a porous medium with CNTS and thermal analysis. *Case Studies in Thermal Engineering*, 12(March), 374–380. https://doi.org/10.1016/j.csite.2018.04.004.

[3] Elnaqeeb, T., Mekheimer, K. S., & Alghamdi, F. (2016). Cu-blood flow model through a catheterized mild stenotic artery with a thrombosis. *Mathematical Biosciences*, 282, 135–146. https://doi.org/10.1016/j.mbs.2016.10.003.
[4] Tang, T. Q., Rooman, M., Vrinceanu, N., Shah, Z., & Alshehri, A. (2022). Blood flow of Au-nanofluid using Sisko model in stenotic artery with porous walls and viscous dissipation effect. *Micromachines*, 13(8), 1–15. https://doi.org/10.3390/mi13081303.
[5] Hu, X., Yin, D., Chen, X., & Xiang, G. (2020). Experimental investigation and mechanism analysis: Effect of nanoparticle size on viscosity of nanofluids. *Journal of Molecular Liquids*, 314, 113604. https://doi.org/10.1016/j.molliq.2020.113604.
[6] Akram, J., Akbar, N. S., & Tripathi, D. (2022). Thermal analysis on MHD flow of ethylene glycol-based BNNTs nanofluids via peristaltically induced electroosmotic pumping in a curved microchannel. *Arabian Journal for Science and Engineering*, 47(6), 7487–7503. https://doi.org/10.1007/s13369-021-06173-7.
[7] Anantha Kumar, K., Sugunamma, V., & Sandeep, N. (2020). Effect of thermal radiation on MHD Casson fluid flow over an exponentially stretching curved sheet. *Journal of Thermal Analysis and Calorimetry*, 140(5), 2377–2385. https://doi.org/10.1007/s10973-019-08977-0.
[8] Khan, U., Shafiq, A., Zaib, A., Sherif, E. S. M., &Baleanu, D. (2020). MHD radiative blood flow embracing gold particles via a slippery sheet through an erratic heat sink/source. *Mathematics*, 8(9), 1597. https://doi.org/10.3390/math8091597.
[9] Saeed, A., Khan, N., Gul, T., Kumam, W., Alghamdi, W., & Kumam, P. (2021). The flow of blood-based hybrid nanofluids with couple stresses by the convergent and divergent channel for the applications of drug delivery. *Molecules*, 26(21), 1–23. https://doi.org/10.3390/molecules26216330.
[10] Ahmed, K., Akbar, T., Muhammad, T., & Alghamdi, M. (2021). Heat transfer characteristics of MHD flow of Williamson nanofluid over an exponential permeable stretching curved surface with variable thermal conductivity. *Case Studies in Thermal Engineering*, 28(October), 101544. https://doi.org/10.1016/j.csite.2021.101544.
[11] McNamara, K., & Tofail, S. A. M. (2017). Nanoparticles in biomedical applications. *Advances in Physics: X*, 2(1), 54–88. https://doi.org/10.1080/23746149.2016.1254570.
[12] Patra, C. R., Bhattacharya, R., Mukhopadhyay, D., & Mukherjee, P. (2010). Fabrication of gold nanoparticles for targeted therapy in pancreatic cancer. *Advanced Drug Delivery Reviews*, 62(3), 346–361. https://doi.org/10.1016/j.addr.2009.11.007.
[13] Seo, W. S., Lee, J. H., Sun, X., Suzuki, Y., Mann, D., Liu, Z., Terashima, M., Yang, P. C., McConnell, M. V., Nishimura, D. G., & Dai, H. (2006). FeCo/graphitic-shell nanocrystals as advanced magnetic-resonance-imaging and near-infrared agents. *Nature Materials*, 5(12), 971–976. https://doi.org/10.1038/nmat1775.
[14] Li, Y., Zhong, M., He, X., Zhang, R., Fu, Y., You, R., Tao, F., Fang, L., Li, Y., & Zhai, Q. (2023). The combined effect of titanium dioxide nanoparticles and cypermethrin on male reproductive toxicity in rats. *Environmental Science and Pollution Research*, 30(9), 22176–22187. https://doi.org/10.1007/s11356-022-23796-x.

[15] Murtaza, M. G., Tzirtzilakis, E. E., & Ferdows, M. (2018). A note on MHD flow and heat transfer over a curved stretching sheet by considering variable thermal conductivity. *World Academy of Science, Engineering and Technology International Journal of Mathematical and Computational Sciences*, 12(2), 38–42 scholar.waset.org/1307-6892/10008601.

[16] Khan, S. A., Hayat, T., & Alsaedi, A. (2020). Entropy optimization in passive and active flow of liquid hydrogen based nanoliquid transport by a curved stretching sheet. *International Communications in Heat and Mass Transfer*, 119, 104890. https://doi.org/10.1016/j.icheatmasstransfer.2020.104890.

[17] Xing, C., Chen, S., Liang, X., Liu, Q., Qu, M., Zou, Q., Li, J., Tan, H., Liu, L., Fan, D., & Zhang, H. (2018). Two-dimensional MXene (Ti3C2)-integrated cellulose hydrogels: Toward smart three-dimensional network nanoplatforms exhibiting light-induced swelling and bimodal photothermal/chemotherapy anticancer activity. *ACS Applied Materials and Interfaces*, 10(33), 27631–27643. https://doi.org/10.1021/acsami.8b08314.

[18] Saharudin, M. S., Ayub, A., Hasbi, S., Muhammad-Sukki, F., Shyha, I., & Inam, F. (2023). Recent advances in MXene composites research, applications and opportunities. *Materials Today: Proceedings*, 2023, 0–4. https://doi.org/10.1016/j.matpr.2023.02.435.

[19] Wang, H., Wu, Y., Zhang, J., Li, G., Huang, H., Zhang, X., & Jiang, Q. (2015). Enhancement of the electrical properties of MXene Ti3C2 nanosheets by post-treatments of alkalization and calcination. *Materials Letters*, 160, 537–540. https://doi.org/10.1016/j.matlet.2015.08.046.

[20] Ansarpour, M., Aslfattahi, N., Mofarahi, M., & Saidur, R. (2022). Numerical study on the convective heat transfer performance of a developed MXene IoNanofluid in a horizontal tube by considering temperature-dependent properties. *Journal of Thermal Analysis and Calorimetry*, 147(21), 12067–12078. https://doi.org/10.1007/s10973-022-11414-4.

[21] Zamhuri, A., Lim, G. P., Ma, N. L., Tee, K. S., & Soon, C. F. (2021). MXene in the lens of biomedical engineering: Synthesis, applications and future outlook. *BioMedical Engineering Online*, 20(1), 1–24. https://doi.org/10.1186/s12938-021-00873-9.

[22] Sarwar, L., Hussain, A., Gamiz, U. F., Akbar, S., & Rehman, A. (2022). Thermal enhancement and numerical solution of blood nanofluid flow through stenotic artery. *Scientific Reports*, 0123456789, 1–11. https://doi.org/10.1038/s41598-022-20267-8.

[23] Dinarvand, S., & Rostami, M. N. (2020). Three-dimensional squeezed flow of aqueous magnetite-graphene oxide hybrid nanofluid: A novel hybridity model with analysis of shape factor effects. *Proceedings of the Institution of Mechanical Engineers, Part E: Journal of Process Mechanical Engineering*, 234(2), 193–205. https://doi.org/10.1177/0954408920906274.

[24] Rathore, N., & Sandeep, N. (2022). Darcy-Forchheimer and Ohmic heating effects on GO-TiO2 suspended cross nanofluid flow through stenosis artery. *The Proceedings of the Institution of Mechanical Engineers, Part C: Journal of Mechanical Engineering Science*, 236, 10470–10485. https://doi.org/10.1177/09544062221105166.

[25] Anantha Kumar, K., Sandeep, N., Sugunamma, V., & Animasaun, I. L. (2020). Effect of irregular heat source/sink on the radiative thin film flow of MHD hybrid ferrofluid. *Journal of Thermal Analysis and Calorimetry*, 139(3), 2145–2153. https://doi.org/10.1007/s10973-019-08628-4.

[26] Hussain, A., Sarwar, L., Rehman, A., Al Mdallal, Q., Almaliki, A. H., & El-Shafay, A. S. (2022). Mathematical analysis of hybrid mediated blood flow in stenosis narrow arteries. *Scientific Reports*, 12(1), 12704. https://doi.org/10.1038/s41598-022-15117-6.

[27] Mabood, F., Shafiq, A., Khan, W. A., & Badruddin, I. A. (2022). MHD and nonlinear thermal radiation effects on hybrid nanofluid past a wedge with heat source and entropy generation. *International Journal of Numerical Methods for Heat and Fluid Flow*, 32(1), 120–137. https://doi.org/10.1108/HFF-10-2020-0636

Chapter 4

Heat transfer analysis in shell-and-tube heat exchanger with helical baffles

Shailandra Kumar Prasad and Mrityunjay Kumar Sinha
National Institute of Technology Jamshedpur

4.1 INTRODUCTION TO THE HEAT-EXCHANGING MACHINE BETWEEN SHELL AND DUCTS

Heat interchangers are broadly utilized in different industrial procedures to transmit heat to separate fluids. One of the most usual and versatile designs of heat interchangers is the shell-and-tube one. They are broadly utilized in applications in which high heat transmit efficacy and firm structure are needed. A tube and shell heat exchanger contains a pile of tubes retained in a cylindrical shell. The shell functions as an accommodation for the tube pile and consists of the fluid that is required to be cooled or heated. The primary design of a tube and shell heat exchanger contains a sequence of parallel tubes scaled into the shell, with baffles situated between the tubes in terms of directing the fluid flow (Tavakoli & Soufivand, 2023). Tube and shell heat interchangers suggest a few benefits, which make them popular in a broad variety of applications.

The configuration of baffles plays an important role in determining the exchange of the thermal energy components of the tube and shell heat exchanger. The composition of helical baffles achieved the most significant attention. Helical baffles are rods or plates in the shape of a spiral, which are installed into the shell to foster tangential or swirling fluid flow. This pattern of swirling flow improves the exchange of thermal energy by enhancing the prolonging and turbulence of the fluid's residence time in the heat exchanger.

This section will concentrate on the exchange of the thermal energy analysis of tube and shell heat interchangers with helical baffles, especially conducting a cross-sectional method (Gugulothu et al., 2023). The effect of helical baffles on the exchange of thermal energy performance, pressure drop, and flow components in the heat interchangers will also be examined.

DOI: 10.1201/9781032712079-4

This study aims to achieve a more precise understanding of the exchange of thermal energy mechanisms and foster the function and operation of tube and shell heat interchangers with helical baffles.

4.2 OVERVIEW OF EXCHANGE OF THE THERMAL ENERGY ANALYSIS

Analysis of exchange of thermal energy is a stream of engineering that concentrates on the research and foretelling of the exchange of thermal energy phenomena within different applications and systems. It includes qualifying and analyzing the transfer of thermal energy among fluids to comprehend heat fluxes, the general performance of the system, and temperature diffusion. This analysis of exchange of the thermal energy is important in various enterprises, incorporating automotive, generation of power, electronics cooling, aerospace, manufacturing procedure, and Heating, Ventilating, and Air Conditioning (HVAC) (Anantha et al., 2022). It assists engineers in developing effective systems of heat exchange, fostering strategies of thermal management and making sure the reliable and secure function of tools.

Three major modes of exchange of the thermal energy are considered by exchange of the thermal energy analysis:

1. **Conduction:** This is the transfer of heat via a concrete material or among solids in immediate connection. It rises due to the collision of particles and vibration of molecules, resulting in the transmission of thermal energy from regions of higher temperature to lower temperature regions.
2. **Convection:** This is the procedure of the exchange of thermal energy that rises in fluids due to the collective impacts of the motion of the fluid and conduction (Wang et al., 2021). It includes the exchange of thermal energy between a moving fluid and a solid surface. It could be neutral or forced, depending on whether the fluid motion is compelled by external forces such as a fan or pump.
3. **Radiation:** This is the transmission of heat via waves of the electromagnet, such as infrared radiation. Not like convention and conduction, radiation could rise within a vacuum and does not need a forum to transfer heat. The appropriate exchange of thermal energy by radiation varies on the temperature of the object, geometrical elements, and properties of the surface.

Computer simulation equipment such as computational fluid dynamics (CFD) and finite element analysis are unusually utilized to crack these calculations and get detailed forecasts of the distribution of temperatures, and heat fluxes. This analysis also believes elements such as boundary conditions, exchange of the thermal energy coefficient, geometric configuration, and material properties.

4.3 TYPES OF BAFFLES IN HEAT-EXCHANGING MACHINE BETWEEN SHELL AND DUCTS

In tube and shell heat interchangers, baffles are utilized to guide the fluid flow and improve the transfer of heat between tube-side and shell-side fluids. Usually, baffles are installed in the shell and are placed among the tubes to make turbulence and enhance the heat transmission surface location. Some commonly utilized kinds of baffles within the tube and shell heat exchange are below:

1. **Segmental Baffles:** These are flat plates that are positioned perpendicular to the tubes, separating the shell into shorter sections (Gu et al., 2022). It extends partially within the diameter of the shell diameter and could have windows to enable some fluid between adjacent baffles.
2. **Disk Baffles:** These are circular plates with slots that are positioned at common intervals with the shell length. It is lined up parallel to the tubes and assists the tube bundle. It assists in managing the flow of fluid and making turbulence, thus enhancing the exchange of thermal energy efficacy.
3. **Rod Baffles:** These are cylindrical rods that are positioned parallelly into the tube bundle. It gives structural assistance to the tubes and supports in guiding the fluid flow. It could have various forms, such as twisted or straight, to foster turbulence and improve exchange of the thermal energy.
4. **Tapered Baffles:** It is utilized to circulate the flow of fluid evenly within the whole tube bundle. It is broader at the side of the inlet and eventually tapers toward the side of the outlet (Biçer et al., 2020). This ensures the distribution of uniform fluid, decreases pressure drop, and improves the efficacy of exchange of the thermal energy.
5. **Helical Baffles:** It is plates or rods of spiral shapes which are placed in the shell. It is designed to generate tangential or swirling fluid flow. It improves exchange of the thermal energy by enhancing the prolonging and turbulence of the living time of fluid in the heat exchanger.

4.4 SIGNIFICANCE OF HELICAL BAFFLES

Helical baffles in heat-exchanging machines between shell and ducts provide many important benefits and help to increase exchange of the thermal energy efficiency. The importance of helical baffles can be seen in the following main reasons:

1. **Improved Heat Transmission:** Helical baffles cause the fluid inside the shell to swirl or flow tangentially, increasing turbulence and improving heat transmission. The fluid's swirling motion improves the exchange of thermal energy efficiency by encouraging better mixing and lengthening (Arani & Uosofvand, 2021). As a result, the fluids on the shell-and-tube side may have smaller temperature differences and higher overall exchange of the thermal energy coefficients.
2. **More Surface Area:** The helical design of the baffles enhances the heat exchanger's effective surface area for exchange of the thermal energy. Compared with straight baffles, the fluid moving along a helical path faces a longer flow path, increasing the surface area open to exchange of the thermal energy. The fluids may exchange heat more readily due to the increased surface area, which enhances the exchange of thermal energy efficiency.
3. **Less Fouling:** In the heat-exchanging machine between shell and ducts, helical baffles help to reduce fouling problems. The whirling flow pattern of helical baffles prevents the growth of stagnant areas and lessens the possibility of deposits or fouling on the exchange of the thermal energy surfaces. Helical baffles help maintain the performance of the heat exchanger and lessen the need for frequent cleaning or maintenance.
4. **Better Fluid Distribution:** Helical baffles help to achieve consistent fluid distribution over the heat exchanger. The swirling action of helical baffles serves to distribute the fluid uniformly throughout the tube bundle, ensuring that each tube has an adequate flow of fluid (Xiao et al., 2020). This even distribution encourages effective heat transmission over the entire exchanger and helps to maximize the usage of the exchange of the thermal energy surface area.
5. **Erosion and Vibration Control:** The tubes are given additional structural support by the helical design of the baffles, which lowers the likelihood of tube vibration and erosion. As stabilizers, the helical baffles reduce the possibility of tube-to-tube or tube-to-baffle contact, which could result in vibrations or damage to the tubes. The heat exchanger's overall stability and lifespan are improved by this feature.

4.5 CROSS-SECTIONAL ANALYSIS OF A HEAT-EXCHANGING MACHINE BETWEEN SHELL AND DUCTS

A heat-exchanging machine between shell and duct's behavior and properties are examined within a particular section as part of a cross-sectional investigation. This research offers insightful information about the heat exchanger's overall performance, fluid flow patterns, pressure drop, and exchange of the thermal energy mechanisms. Here are some crucial factors a cross-sectional analysis takes into account:

1. **Temperature Distribution:** The analysis of the cross-section of the heat exchanger enables the assessment of temperature fluctuations throughout the heat exchanger (Arani & Uosofvand, 2021). It aids in locating hot and cold places as well as temperature gradients and high and low temperatures throughout the exchanger. As per Figure 4.1, evaluating the exchange of thermal energy effectiveness and making sure that the desired thermal criteria are satisfied, it is essential to understand the temperature distribution.
2. **Fluid Flow Patterns:** Engineers can evaluate the fluid flow patterns on the shell side and tube side by examining the cross-section. This entails examining the flow rates, directionality, and any potential recirculation or dead zones. It is possible to spot places with poor fluid mixing, high or low flow rates, and probable hot spots by understanding the flow patterns.
3. **Exchange of the Thermal Energy Surface Area:** The cross-sectional examination enables the assessment of the heat exchanger's actual exchange of the thermal energy surface area. As per Figure 4.2, engineers can analyze the potential for effective exchange of thermal energy between the fluids by calculating the surface area which is available for heat exchange. To maximize the exchange of the thermal

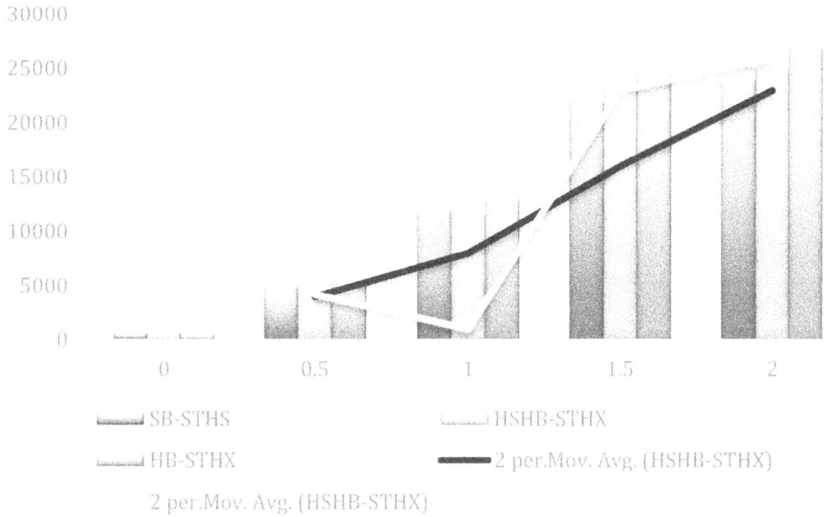

Figure 4.1 Shell diameter against heat transfer coefficient for shell side with three types of headers and triangular pitch.

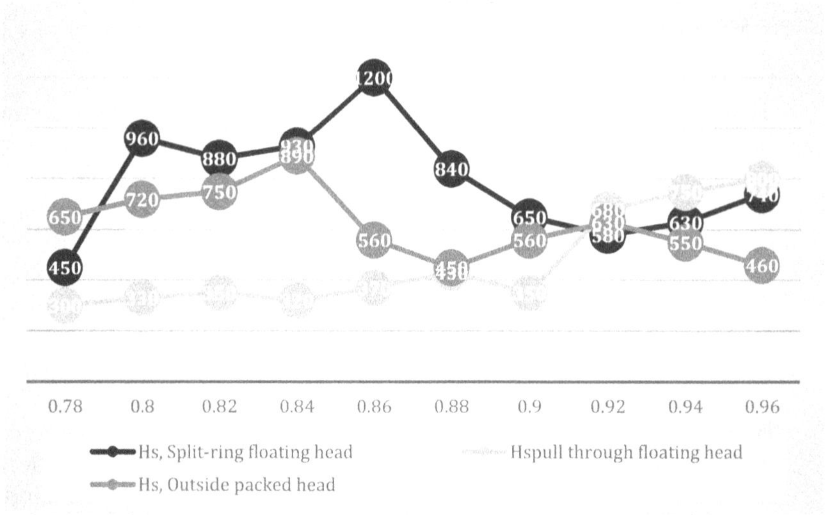

Figure 4.2 Total heat transfer coefficient changing with tube-side flow rate while keeping the shell-side flow rate constant.

energy, the design of the surface area, and configuration of the heat exchanger are optimized with the aid of this analysis.

4. **Pressure Drop:** A cross-sectional analysis should include a pressure drop analysis. Analyzing the variations in fluid pressure as the fluid passes through the heat exchanger is required. Engineers can evaluate flow resistance, spot locations of significant pressure drops, and optimize the heat exchanger design to reduce pressure losses by analyzing the pressure drop over the cross-section.
5. **Fluid Velocity and Residence Time:** Information on fluid velocities and residence times inside the heat exchanger is provided by the cross-sectional study. Engineers can evaluate the flow rates and distribution of the fluids over the cross-section by knowing the fluid velocities (Saffarian et al., 2019). Analysis of the fluid's residence time inside the heat exchanger, which affects the effectiveness of exchange of the thermal energy and overall performance, can be done.
6. **Baffle Effectiveness:** The cross-sectional study enables the assessment of baffle performance in controlling fluid flow and improving heat transmission. Engineers can evaluate the effects of various baffle configurations on flow turbulence, exchange of thermal energy enhancement, and pressure drop by studying the behavior of the fluid flow around the baffles.

A heat-exchanging machine between shell and duct's performance may be assessed, its design can be optimized, and possible areas for improvement can be found using cross-sectional analysis. It assists engineers in making knowledgeable decisions about fluid flow characteristics, system performance overall, and exchange of thermal energy efficiency, resulting in more effective and efficient heat exchange operations.

4.6 GEOMETRY AND DESIGN CONSIDERATIONS

A heat-exchanging machine between shell and duct's performance and efficiency are greatly influenced by its geometry and design. Here are some important things to think about:

1. **Tube Geometry:** The geometry of the tubes must be taken into account during design. The effectiveness of the exchange of thermal energy and pressure drop is influenced by variables, for example, energy rates are typically offered by larger diameter tubes, but they may also increase the heat exchanger's overall size and price. The accessible exchange of the thermal energy surface area is influenced by tube length, and flow distribution and pressure drop characteristics are influenced by arrangement.
2. **Shell Geometry:** The diameter, length, and shape of the shell all fall under its geometry. The heat exchanger's overall dimensions and capacity are determined by the shell diameter as shown in Figure 4.3. The number of tube passes and the fluids' residence duration inside the exchanger are both influenced by the shell length (Ahmad & Mahmud, 2022). Depending on the particular use and available space, either a cylindrical or rectangular shell is utilized.

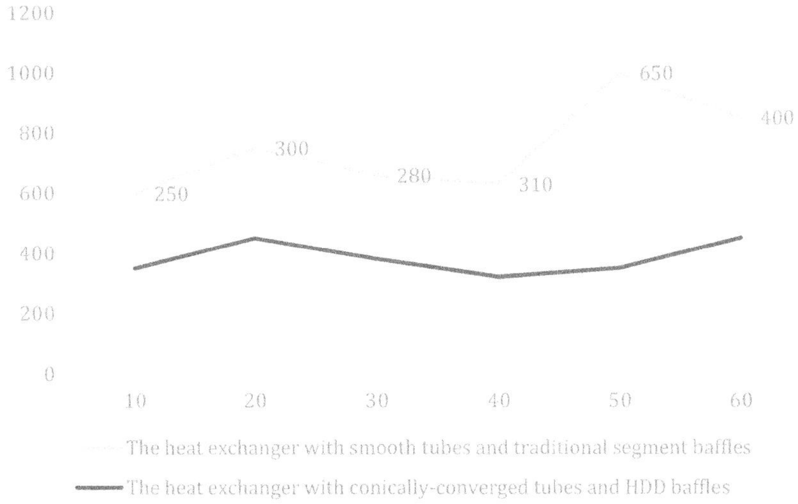

Figure 4.3 Tube diameter, length, and arrangement (in-line or staggered).

3. **Design of the Baffles:** Baffles are essential for controlling fluid flow and increasing heat transmission. The design of baffles takes into account factors such as the type of baffles (segmental, helical, disc, etc.), the distance between baffles, the cut of the baffles, and the thickness of the baffles. These design elements have an impact on the heat exchanger's flow distribution, fluid mixing, pressure drop, and exchange of thermal energy efficiency.
4. **Tube Layout:** The flow distribution and heat transmission properties of the shell are influenced by the arrangement of the tubes. Square pitch, triangular pitch, rotated square, and rotated triangular are typical tube configurations. Regarding pressure drop, exchange of thermal energy efficiency, and simplicity of cleaning and maintenance, each configuration offers benefits and drawbacks.
5. **Pass Partitioning:** Pass partitioning is the process of employing divider plates or partitions to divide the shell-side flow into various passes or portions. Pass partitioning enhances exchange of the thermal energy efficiency, lowers pressure drop, and provides better fluid flow management. In addition, it aids in achieving thermal and fluid equilibrium throughout the heat exchanger.

4.7 FLUID FLOW AND EXCHANGE OF THE THERMAL ENERGY MECHANISMS

Fundamental features of a heat-exchanging machine between shell and ducts include fluid flow and exchange of the thermal energy processes. Optimizing the heat exchanger's design and performance requires a thorough understanding of these mechanisms. The main heat transmission and fluid flow mechanisms at play are as follows:

4.7.1 Flow mechanisms for fluids

1. **Crossflow:** In a heat-exchanging machine between shell and ducts, the fluid on the shell side flows across the exterior of the tubes while the fluid on the tube side flows inside the tubes. A crossflow pattern is created by this arrangement. By causing fluid mixing and increasing fluid contact, crossflow encourages effective heat transmission.
2. **Parallel Flow:** In a parallel flow configuration, both the fluids flowing into and out of the tubes and shells flow in the same direction (Pasupuleti et al., 2021). As of the bigger temperature difference between the fluids in this design at the inlet, there is a higher rate

of overall exchange of thermal energy. However, as the exchanger's length increases, the temperature differential reduces.

3. **Counter-Flow:** In a counter-flow configuration, the fluids flowing in the tube and the shell move in opposing directions. The most effective exchange of thermal energy is provided by this setup. The temperature differential between the fluids remains very high along the whole length of the exchanger as they move against one another, enhancing heat transmission.

4.7.2 Mechanisms for exchange of the thermal energy

1. **Conduction:** Conduction is the process by which heat is transferred between solid objects in direct contact or within solid materials. In a heat-exchanging machine between shell and ducts, heat is transferred from the hot fluid to the cold fluid through the tube walls. The rate of conduction is influenced by the materials' thermal conductivity and temperature gradient.
2. **Convection:** The heat transfer method that combines the effects of conduction and fluid motion is known as convection. Convective exchange of the thermal energy occurs on both the tube side and the shell side of the heat-exchanging machine between shell and ducts. Convection is the process by which heat is transferred from the fluid flowing inside the tubes or across the tubes (tube side). Temperature gradients, fluid viscosity, and fluid velocity all have an impact on the convective exchange of thermal energy.
3. **Radiation:** Radiation is an electromagnetic wave-based method of transferring heat. In comparison to conduction and convection, radiation exchange of thermal energy normally has little effect in heat-exchanging machine between shell and ducts (Mazdak et al., 2023). Radiation can, however, play a bigger role in high-temperature applications, such as furnaces or other industrial operations. Heat transport through radiation is influenced by the surfaces' temperature and emissivity.

The performance and total exchange of the thermal energy efficiency of the heat-exchanging machine between shell and ducts are determined by the interaction of these exchanges of the thermal energy mechanisms and the fluid flow patterns. Engineers can improve design parameters, such as tube size, flow rates, baffle designs, and fluid characteristics to ensure an effective exchange of thermal energy and satisfy particular application needs by comprehending these mechanisms.

4.8 THERMAL PERFORMANCE EVALUATION

Thermal performance evaluation is a vital step to assess the effectiveness and efficiency of a heat-exchanging machine between shell and ducts. It includes the analysis of different parameters and the calculations to ensure the process of exchange of the thermal energy efficiency to perform the shell and heat generator tube (Gu et al., 2022). After having the proper methods of calculating the thermal nature of the evolution of the system are given below:

1. **Overall Exchange of Thermal Energy:** The coefficient of overall exchange of thermal energy presents the combination of the resistance of the thermal factors of both the shell-side and tube-side fluids. This is quite responsible for the convection, radiation, and conduction of the mechanisms of heat.
2. **Log Mean Temperature Difference (LMTD):** Another vital and sensitive parameter to be used in the determination of the temperature force to drive the exchange of thermal energy in a tube and shell to exchange the heat. It is calculated in Figure 4.4, according to the temperature gap between the cold and the hot fluids at different points with the exchanger.
3. **Rate of Exchange of the Thermal Energy:** Another quantifying factor of heat exchange is the amount of energy transferred to transfer the energy between the cold and hot fluids inside the heat exchanger. The calculation is used for the analysis of the product of the gross coefficient of exchange of the thermal energy, the surface area of

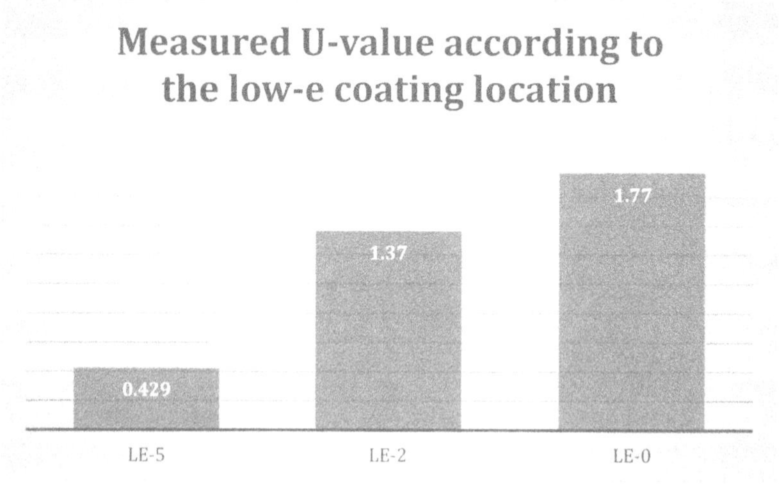

Figure 4.4 The measured U-value according to the Low-E coating location.

exchange of the thermal energy, and the difference in temperature between the fluids.

4. **Effectiveness:** The effectiveness of the heat exchanger can be measured with the help of the ratio of the actual exchange of thermal energy to the most extreme method. Which can be influenced by the perimeters of the configuration of the flow, heat-exchanging geometry, and the property of fluids. The indicator of efficiency of the heat exchanger is mostly dependent on the rate of transfer of thermal energy.
5. **Pressure Drop:** Pressure drop refers to the loss of pressure of fluid because it flows through the heat exchanger. The evaluation of the decrease of pressure is quite essential in the determination of the energy is quite necessary for overcoming the resistance of flow to resist with the exchanger, increasing the cost of pumping, and the reduction of the efficiency of the system.
7. **Efficiency:** The effective measures to be taken in getting the desired outcome in the heat-exchanging process (Desai et al., 2021). The dependent factors to achieve the rate of exchange of the thermal energy, input energy, and the drop of pressure.

The growth of the thermal performance permits the engineers to assess the optimization of the design, and adoption of the parameters, to find out the more efficient indicator to enhance improvement, by making the decisions to be informed related to the modification of the system, by ensuring the optimal transfer of heat efficiently to perform.

4.9 EXPERIMENTAL TECHNIQUES AND DATA COLLECTION

The techniques of collection of data to the necessary components to evaluate the heat-exchanging machine between shell and ducts. The involvement to be conducted the research in measuring the chief parameters of gathering the data that are used in validating the theory-based models, regulating the designs, and measuring the performance of the heat exchanger. Some of the commonly used techniques of experimental techniques as well as the methods of collecting data are given below:

1. **Measurement of Temperature:** Temperature measurement in analyzing and evaluating the heat-exchanging machine between shell and ducts needs to be evaluated. This includes the experiment in measuring the criteria to accumulate the data to be utilized in the theoretical models, and to the inlet and outlet temperature of fluids, in several temperature differences.

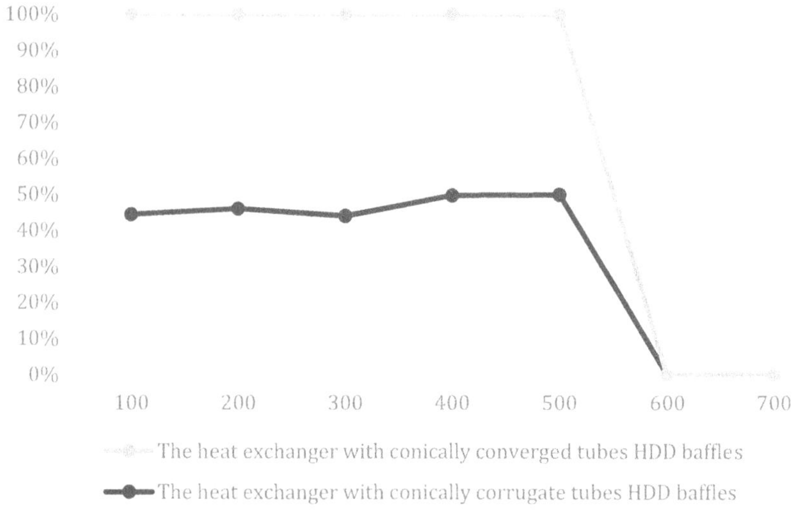

Figure 4.5 Shell-and-tube heat exchangers performance improvement employs.

2. **Managing Rate of Flow:** Flow rates of the fluids are required to be measured accurately which is vital to evaluate the performance of heat exchange. The flow mattress, such as electromagnetic, ultrasonic, or differential flow meters measuring pressure, can be used while measuring the rate of flow on the shell side and tube side. The differences are clearly visible in Figure 4.5.
3. **Pressure Measurements:** Pressure measurements are necessary to examine the fluid flow characteristics and gauge the pressure drop across the heat exchanger. At various points within the system, such as the entrance and outflow of the shell-and-tube sidewalls, pressures are measured using pressure sensors or gauges.
4. **Measurements of the Exchange of the Thermal Energy Rate:** Determining the exchange of the thermal energy rate is essential for assessing the heat exchanger's thermal performance. Energy balance formulae based on fluid flow rates and temperature readings can be used to calculate the indirect exchange of the thermal energy rate (Liu et al., 2020). Alternatively, calorimetric techniques or heat balancing tests can be used to determine the rate of heat transmission.
5. **Measurements of Heat Exchanger Efficacy:** Using inlet and outlet temperature readings, flow rates, and energy balance calculations, it is feasible to experimentally determine a heat exchanger's efficacy, which compares the actual exchange of the thermal energy to the maximum amount of exchange of the thermal energy.
6. **Fouling Measurements:** To assess the effects of fouling on heat exchanger performance, fouling measurement techniques are used.

Fouling can be assessed either directly by tracking the deposit thickness or inferentially by analyzing the degradation in heat transmission over time. To evaluate the fouling characteristics, techniques such as heat resistances or cleanability tests are performed.

4.10 NUMERICAL SIMULATION AND COMPUTATIONAL FLUID DYNAMICS (CFD)

Heat-exchanging machine between shell and duct performance is analyzed and optimized using CFD and numerical simulation. An overview of numerical simulation and CFD concerning heat-exchanging machine between shell and ducts is provided below:

1. **Mathematical Modeling:** Mathematical models that represent the physical behavior of fluid flow and exchange of the thermal energy within the heat exchanger are created through numerical simulation (Silva, 2022). These models are based on fundamental ideas such as constitutive relations (fluid characteristics, turbulence models) and conservation equations (mass, momentum, and energy).
2. **Computational Fluid Dynamics (CFD):** CFD is a specialized subset of numerical simulation that is geared toward modeling and simulating fluid flow and heat transport. It entails applying numerical methods to solve the governing equations iteratively after discretizing them using computational grids or meshes.
3. **Geometry and Meshing:** A CFD program is used to model the geometry of the heat exchanger, which includes the tubes, shell, baffles, and inlet/outlet connections. Mesh techniques are then used to separate the geometry into discrete elements or cells. The accuracy and effectiveness of the simulation are influenced by the mesh quality, including cell size and distribution.
4. **Boundary Conditions:** In the CFD simulation, boundary conditions specify the inflow, outflow, and wall conditions. The flow rates, temperatures, and characteristics of the fluids entering and leaving the heat exchanger are specified by the circumstances at the inlet and outlet (Hajabdollahi et al., 2013). The temperatures, roughness, and exchange of the thermal energy coefficients of the walls are considered to be the wall conditions.
5. **Turbulence Modeling:** The exchange of the thermal energy and fluid flow within the heat exchanger are significantly influenced by turbulence, a complicated and highly nonlinear process. To capture and anticipate the turbulent flow behavior, a variety of turbulence models, including Reynolds-Averaged Navier-Stokes (RANS) models and Large Eddy Simulation (LES), can be used.

6. **Heat Transport Analysis:** CFD models offer precise forecasts of the properties of heat transport inside the heat exchanger (Gugulothu et al., 2021). These consist of the exchange of thermal energy coefficients, temperature distributions, and overall exchange of the thermal energy rates. Engineers can use CFD to examine how various design factors, such as tube diameter, baffle arrangement, or flow rates, affect exchange of the thermal energy efficiency.
7. **Validation and Optimization:** To confirm the correctness and dependability of the simulations, CFD results are often compared with experimental data. To enhance the exchange of thermal energy performance and reduce energy consumption, engineers can investigate numerous design possibilities, assess the effects of various operating situations, and choose the most effective configurations.

Engineers may learn a lot about the flow patterns, exchange of the thermal energy processes, and overall performance of heat-exchanging machine between shell and ducts by using numerical simulation and CFD (Abdelkader et al., 2020). Engineers may learn a lot about the flow patterns, exchange of the thermal energy processes, and overall performance of heat-exchanging machine between shell and ducts by using numerical simulation and CFD.

4.11 EXCHANGE OF THE THERMAL ENERGY ENHANCEMENT WITH HELICAL BAFFLES

Helix-shaped baffles are a widely researched and used method for improving heat transmission in heat-exchanging machines between shell and ducts. With their spiral-shaped design, helical baffles have many benefits for enhancing heat transmission efficiency. The following are some significant ways that helical baffles improve the exchange of thermal energy:

1. **Increased Turbulence:** The helical design of the baffles causes the fluid inside the heat exchanger to swirl or flow in a tangential direction. This swirling action increases turbulence, which improves fluid mixing and facilitates better heat transmission.
2. **Prolonged Residence Duration:** The helical baffles' swirling flow pattern lengthens the fluid's duration in residence inside the heat exchanger (Saha & Hossain, 2021). The fluid's long journey and longer time in contact with the exchange of the thermal energy surfaces are made possible by the helical trajectory.
3. **Increased Exchange of the Thermal Energy Coefficient:** On both the shell side and tube side of the exchanger, the turbulence produced by the helical baffles greatly raises the exchange of thermal energy coefficient. A greater rate of exchange of the thermal energy between the

fluids is caused by the improved exchange of the thermal energy coefficient, improving the overall exchange of thermal energy efficiency.

4. **Applications for Viscous Fluids:** Helical baffles are especially useful in improving heat transmission in applications for viscous fluids. The helical baffles create a whirling flow that facilitates the breakdown of boundary layers and enhances exchange of the thermal energy efficiency in very viscous fluids (Chen et al., 2020). As of this, helical baffles are excellent choices for processes involving high viscosities in industry, such as oil cooling.

4.12 COMPARISON WITH CONVENTIONAL BAFFLE CONFIGURATIONS

There are various considerations when comparing helical baffles with traditional baffle configurations in heat-exchanging machines between shell and ducts. The following is a comparison between helical baffles and regular baffle arrangements:

1. **Improved Heat Transmission:** With the increasing turbulence and extending the fluid's residence time, helical baffles are noted for improving heat transmission. Improvements in the exchange of thermal energy coefficients and total exchange of the thermal energy performance result from this (Mehrjardi et al., 2023). Compared with helical baffles, conventional baffle shapes such as segmental or disc baffles may not be as successful at generating swirling flow and maximizing the exchange of thermal energy rates. Nevertheless, they do generate some level of turbulence and exchange of thermal energy increase.
2. **Fluid Mixing:** Due to the swirling flow pattern they produce, helical baffles encourage better fluid mixing. By guaranteeing a more even distribution of temperatures throughout the heat exchanger, this improved mixing raises the effectiveness of the exchange of thermal energy. When compared with the whirling flow produced by helical baffles, the effect of conventional baffle layouts may be restricted.
3. **Pressure Drop:** When designing and using heat interchangers, pressure drop is a crucial factor. Due to the increased turbulence and flow resistance, they produce, helical baffles typically cause a bigger pressure drop (Gu et al., 2023). Conventional baffle layouts, on the other hand, might offer a reduced pressure drop, particularly if the baffles are placed further apart. It's crucial to remember that when choosing the right baffle layout, system limits as well as the pressure drop requirements of a particular application must be taken into account.
4. **Fouling and Cleaning:** By avoiding stagnant areas and encouraging greater fluid flow, helical baffles typically lessen fouling. Conventional

baffle configurations may also help reduce fouling to some extent, but their effectiveness may vary depending on the flow patterns and cleaning accessibility within the heat exchanger.

5. **Design Freedom:** Helical baffles provide for adaptation to specific exchange of thermal energy requirements by allowing for design freedom in terms of the number of turns, pitch, and spacing. On the other hand, conventional baffle arrangements have more standardized designs and could have restrictions on design flexibility.
6. **Applications for Viscous Fluids:** Helical baffles are especially useful in heat interchangers for viscous fluids. Helicopter-shaped baffles create a whirling flow pattern that improves exchange of the thermal energy in viscous fluids and breaks up boundary layers. In situations involving viscous fluids, conventional baffle arrangements could not offer the same level of exchange of thermal energy increase.

4.13 CHALLENGES AND LIMITATIONS

While heat-exchanging machines between shell and ducts provide several advantages, they have come with some challenges and setbacks. Some of the common challenges and limitations have to be associated with heat-exchanging machines between shell and ducts:

1. **Size and Space Requirements:** Heat-exchanging machines between shell and ducts have to be typically maintained in comparison to the size of other types of heat (Debnath & Pradhan, 2023). This can be quite helpful in fulfilling the requirements e respective of time and space especially since the complete installation of applications can reduce the availability of space.
2. **Pressure Drop:** Heat-exchanging machines between shell and ducts can produce a high drop in pressure, especially inside the cell. More preset drops need higher energy to plump the fluids throughout the system which can have an overall effect for making the system more efficient and reduce operational cost.
3. **Fouling and Maintenance:** Heat-exchanging machines between shell and ducts are quite effective in fouling to accumulate the depositions in the exchange of the thermal energy surface (Khan et al., 2023). The transfer efficiency can be reduced through fouling. This specific especially challenges the applications involving fluids to the potential fluid composition in a very complex manner.
4. **Temperature Cross-Contamination:** In a few cases, tube heat interchangers experience cross-contamination of temperature between tube-side fluid and cell-side fluids. Please can be seen at the time of mixing all leakage between the different fluid streams to the

unnecessary exchange of the thermal energy between them. This required work.

5. **Limited Exchange of Thermal Energy:** Helical battles enhance the transfer of heat in silent heat interchangers, and several obstacles can be found to achieve the optimal exchange of the thermal energy ability (Saha & Hossain, 2021). Flow rates and properties of fluid are the main constraints that show a higher impact in enhancing the degree of exchange of thermal energy.
6. **Flow Mal-distribution:** Improper distribution of fluid can provide different states of flow throughout the tubes in heat-exchanging machines between shell and ducts which is the main cause of the reduction of exchange of thermal energy effectively. Slow heat can produce unexpected results to enhance blockage inside the tubes.
7. **Limited Thermal Response:** The limited thermal response can be achieved through the heat-exchanging machine between shell and ducts as a large volume of fluid can be produced inside the system. The heat load or condition of fluid may be quite slow. These put an obstacle to applying the dynamic operation in thermal response.
8. **Cost:** In compact to the other heat interchangers heat-exchanging machines between shell and ducts are quite precious these can limit the rapid thermal response operation, for their higher cost.

4.14 FUTURE TRENDS AND RESEARCH DIRECTIONS

With the advancement of technology, so many future trends will emerge in the sector of shell-and-tube heat exchanger (Heydari et al., 2018). The trends are oriented to achieve the aim and improvisation of the performance, by the enhancement of operation and design of the heat exchanger. Some of the specific areas are given below:

1. **Advanced Materials:** While exploring the building of advanced materials that may increase thermal conductivity, and reduce resistance, to construct the exchange of heat (Arumsari & Ginting, 2023). Nanostructured coverings, graphene, and alloy metals have to be the substance of advanced research, to achieve the desired outcome to innovate the materials while transferring efficiency and strength.
2. **Compact Designs:** The trend to the compact design is mainly dependent on lessening the footprint of the heat-exchanging machine between shell and ducts irrespective of adjusting the performance. To optimize the header design and arrangement of type, baffling of the configuration has to be implied to meet the required rate of exchange of the thermal energy to reduce the pressure drops in reducing physical dimensions.

3. **Enhancement of Techniques of Exchange of the Thermal Energy:** Advanced research is being carried out to explore and develop the exchange of thermal energy techniques for the necessary improvement of the performance of the innovation to baffle the design and the augmentation methods have been improved to improve the surface coatings and the optimization of patterns of flow.
4. **Methods of Computing:** Computational methods have been scaling up day by day (Tuncer et al., 2023). In developing heat interchangers to reduce energy consumption, and its impact on the environment, several kinds of research have been conducted.

4.15 CONCLUSION

To conclude, heat-exchanging machines between shell and ducts are often used in different industries for easy transfer between fluids. They have to offer advantages, such as design flexibility, strong construction, and rates of high exchange of thermal energy. Thus, they also present limitations and challenges that need careful consideration during function, maintenance, and design. To overcome these issues and further improve the performance of heat interchangers, the current study concentrates on different elements. These incorporate the growth of advanced materials, examination of micro-scale heat interchangers, advancement of heat exchanger methods, optimization of compact designs, enhancement of computational techniques, integration of smart technologies, energy efficacy, and sustainability. Future improvements in these rooms hold the possibility to enhance the efficacy of exchange of the thermal energy, decrease energy consumption, mitigate fouling, and enhance the designs of the heat exchanger. This would donate to more efficient and sustainable procedures of industries.

REFERENCES

Abdelkader, B. A., Jamil, M. A., & Zubair, S. M. (2020). Thermal-hydraulic characteristics of helical baffle shell-and-tube heat exchangers. *Heat Transfer Engineering*, 41(13), 1143–1155.

Ahmad, F., & Mahmud, S. (2022). Nanofluid Implementation in Heat Exchanger Geometries (Doctoral dissertation, Department of Mechanical and Production Engineering (MPE), Islamic University of Technology (IUT)).

Anantha, S., Gnanamani, S., Mahendran, V., Rathinavelu, V., Rajagopal, R., Tafesse, D., & Nadarajan, P. (2022). A CFD investigation and heat transfer augmentation of double pipe heat exchanger by employing helical baffles on shell and tube side. *Thermal Science*, 26(2 Part A), 991–998.

Arani, A. A. A., & Uosofvand, H. (2021). Double-pass shell-and-tube heat exchanger performance enhancement with new combined baffle and elliptical tube bundle arrangement. *International Journal of Thermal Sciences*, 167, 106999.

Arumsari, A. G., & Ginting, P. J. (2023). Analysis of heat transfer coefficient of shell and tube on heat exchanger using Heat Transfer Research Inch (HTRI) software. *Formosa Journal of Sustainable Research*, 2(5), 1175–1184.

Biçer, N., Engin, T., Yaşar, H., Büyükkaya, E., Aydın, A., & Topuz, A. (2020). Design optimization of a shell-and-tube heat exchanger with novel three-zonal baffle by using CFD and taguchi method. *International Journal of Thermal Sciences*, 155, 106417.

Chen, Y., Zhu, Z., Wu, J., Zheng, S., Song, N., & Su, J. (2020). Mass center equivalent rectangle model for universal correlation of hydrothermal features of helical baffle heat exchangers. *Applied Thermal Engineering*, 174, 115307.

Debnath, P., & Pradhan, M. (2023). A recent state of art review on heat transfer enhanced characteristics and material selection of SCTHX. *Proceedings of the Institution of Mechanical Engineers, Part E: Journal of Process Mechanical Engineering*, 237(3), 1005–1013.

Desai, K., Rathod, P., Bariya, P., Doshi, R., & Bagul, A. (2021). Experimental analysis and investigation on twisted tube shell type heat exchanger. *Research Journal of Engineering Sciences*, 10(3), 36–38.

Gu, X., Zheng, Z., Xiong, X., Jiang, E., Wang, T., & Zhang, D. (2022). Heat transfer and flow resistance characteristics of helical baffle heat exchangers with twisted oval tube. *Journal of Thermal Science*, 31(2), 370–378.

Gu, X., Shi, Q., Gao, W., Li, M., & Wang, D. (2023). Performance analysis and structural optimization of torsional flow heat exchangers with sinusoidal corrugated baffle. *Journal of Thermal Science*, 32(2), 680–691.

Gugulothu, R., Sanke, N., Ahmed, F., & Kumari Jilugu, R. (2021). Numerical study on shell and tube heat exchanger with segmental baffle. In *Proceedings of International Joint Conference on Advances in Computational Intelligence: IJCACI 2020* (pp. 309–318). Springer Singapore

Hajabdollahi, Z., Hajabdollahi, F., Tehrani, M., & Hajabdollahi, H. (2013). Thermo-economic environmental optimization of Organic Rankine Cycle for diesel waste heat recovery. *Energy*, 63, 142–151.

Heydari, A., Shateri, M., & Sanjari, S. (2018). Numerical analysis of a small size baffled shell-and-tube heat exchanger using different nano-fluids. *Heat Transfer Engineering*, 39(2), 141–153.

Khan, Y., Raman, R., Rashidi, M. M., Caliskan, H., Chauhan, M. K., & Chauhan, A. K. (2023). Thermodynamic analysis and experimental investigation of the water spray cooling of photovoltaic solar panels. *Journal of Thermal Analysis and Calorimetry*, 148(12), 5591–5602.

Liu, L., Shen, T., Zhang, L., Peng, H., Zhang, S., Xu, W., Bo, S., Qu, S., & Ni, X. (2020). Experimental and numerical investigation on shell-and-tube exhaust gas recirculation cooler with different tube bundles. *Heat and Mass Transfer*, 56, 601–615.

Mazdak, S., Sheikhzadeh, G. A., & Fattahi, A. (2023). Numerical analysis of a heat exchanger with curved segmental baffle and Cassini oval cross-section tubes in various bundle arrangements. *Journal of Thermal Analysis and Calorimetry*, 148, 1–18.

Mehrjardi, S. A. A., Khademi, A., Said, Z., Ushak, S., & Chamkha, A. J. (2023). Effect of elliptical dimples on heat transfer performance in a shell and tube heat exchanger. *Heat and Mass Transfer*, 59, 1–11.

Pasupuleti, R. K., Bedhapudi, M., Jonnala, S. R., & Kandimalla, A. R. (2021). Computational analysis of conventional and helical finned shell and tube heat exchanger using ANSYS-CFD. *International Journal of Heat & Technology*, 39(6), 1755–1762.

Saffarian, M. R., Fazelpour, F., & Sham, M. (2019). Numerical study of shell and tube heat exchanger with different cross-section tubes and combined tubes. *International Journal of Energy and Environmental Engineering*, 10, 33–46.

Saha, B., & Hossain, M. T. (2021). Impact of baffle spaces and baffle numbers on pressure drop and heat transfer in a shell and tube heat exchanger: a CFD analysis. *Advancement in Mechanical Engineering and Technology*, 4(2, 3), 26–39.

Silva, R. A. D. A. (2022). Shell and Tube Heat Exchanger with Helical Baffles and Graphene Nanofluid: A CFD Analysis of Its Performance and Characteristics (Master's thesis, Universidade Federal de Pernambuco).

Tavakoli, M., & Soufivand, M. R. (2023). Performance evaluation criteria and entropy generation of hybrid nanofluid in a shell-and-tube heat exchanger with two different types of cross-sectional baffles. *Engineering Analysis with Boundary Elements*, 150, 272–284.

Tuncer, A. D., Khanlari, A., Sözen, A., Gürbüz, E. Y., & Variyenli, H. I. (2023). Upgrading the performance of shell and helically coiled heat exchangers with new flow path by using TiO_2/water and CuO-TiO_2/water nanofluids. *International Journal of Thermal Sciences*, 183, 107831.

Wang, J., Bian, H., Cao, X., & Ding, M. (2021). Numerical performance analysis of a novel shell-and-tube oil cooler with wire-wound and crescent baffles. *Applied Thermal Engineering*, 184, 116298.

Xiao, J., Wang, S., Ye, S., Wen, J., & Zhang, Z. (2020). Multiphysics field coupling simulation for shell-and-tube heat exchangers with different baffles. *Numerical Heat Transfer, Part A: Applications*, 77(3), 266–283.

Chapter 5

Residual and computational time analysis of the boundary-layer flow

Atul Kumar Ray
Madhav Institute of Technology and Science Gwalior

B. Vasu
Motilal Nehru National Institute of Technology Allahabad

Amit Kumar
University of Petroleum & Energy Studies, Dehradun

Dig Vijay Tanwar
Graphic Era (Deemed to be University), Dehradun

Divya Chaturvedi and Minakshi Poonia
Madhav Institute of Technology and Science Gwalior

5.1 INTRODUCTION: BACKGROUND AND DRIVING FORCES

The viscous flow over stretching surface has extensive applications in biomedical, food, geophysical fluid dynamics and polymer industry [1]. The concept of a boundary layer for stretching surface was pioneered by Sakidis [2]. The study on the behaviour of boundary-layer flow of quiescent fluid over a flat surface was done by Crane [3]. In the beginning of the 20th century, Liao [4] provided homotopy series solutions to the non-linear, boundary-layer problems. Recently, Ray [5] studied the flow of non-Newtonian Jeffrey nanofluid due to a permeable vertical sheet with Cattaneo–Christov heat flux model.

HAM [6] is a robust method for providing semi-analytic solutions to complex non-linear coupled differential equations. HAM has been applied to many problems from different fields of engineering and applied sciences. The fortunate thing in HAM is that it does not depend on any perturb parameters and guarantees the convergence of homotopy series. Liao [7]

DOI: 10.1201/9781032712079-5

explained the process in the homotopy approach to solve the non-linear differential equations. Ray et al. [8] studied the flow of Spriggs fluid past an impulsive motion of a flat surface using HAM. Recently, Be'g et al. [9] implemented HAM to the micropolar fluid flow problem as an application in polymer coating.

The motivation of the present investigation is to explore the significance of residual error, computational time and convergence-control parameters in HAM for viscous fluid flow past a permeable surface [10]. The mathematical modelling is explained in Section 5.2. The solution process using HAM is provided in Sections 5.3, and 5.4 has the result and analysis of the study.

5.2 MATHEMATICAL FORMULATION

Consider 2-D viscous flow in the first quadrant $(X, Y>0)$ where (X, Y) represents the Cartesian coordinate system and flow originates by motion of an impermeable plate at $Y = 0$ in its plane. The viscous fluid is incompressible. Let $U_w(X) = A(X+B)^k$ represent the speed of a flat plate where constants A and B are positive. The governing equation for the flow is given by

$$\frac{\partial U}{\partial X} + \frac{\partial V}{\partial Y} = 0 \tag{5.1}$$

$$U\frac{\partial U}{\partial X} + V\frac{\partial U}{\partial Y} = \nu\frac{\partial^2 U}{\partial Y^2} \tag{5.2}$$

Subject to boundary

$$\text{At} \quad Y = 0, \quad U = U_w(X), \quad V = 0 \tag{5.3}$$

$$\text{At} \quad Y \to \infty, \quad U = 0 \tag{5.4}$$

where U and V are velocity component in, respectively, X and Y directions. ϑ is the kinematic viscosity. Defining stream function χ as $U = \frac{\partial \chi}{\partial y}$ and $V = -\frac{\partial \chi}{\partial x}$. Using similarity transformation, considering χ as

$$\chi = \sqrt{A\nu}(X+B)f(\eta), \quad \eta = \sqrt{\frac{a}{\nu}}Y(X+B)^{\frac{k-1}{2}} \tag{5.5}$$

by using equation (5.5)

$$U = A(X+B)^k f'(\eta) \text{ and } V = -\left[\sqrt{A\nu}\frac{k+1}{2}(X+B)^{\frac{k-1}{2}}f(\eta) + \frac{k-1}{2}A(X+B)^{k-1}Yf'(\eta)\right]$$

The equation (5.2) reduced to

$$f'''(\eta)+\frac{k+1}{2}f(\eta)f''(\eta)-k\left(f'(\eta)\right)^2=0 \tag{5.6}$$

boundary condition (5.3) and (5.4) becomes

$$f(0)=0, \quad f'(0)=1, \quad f'(\infty)=0 \tag{5.7}$$

5.3 SOLUTION PROCESS

The frame of HAM includes the auxiliary linear operator $\mathcal{L}$, initial guess $f_0(\eta)$ and especially the convergence-control parameter c_0 which will form equations for zeroth and mth order of deformation. Choosing the initial approximation as

$$f_0(\eta)=\sigma^*+(1-2\sigma^*)e^{-\eta}-(1-\sigma^*)e^{-2\eta} \tag{5.8}$$

where $\sigma^*=f_0(\infty)$ is unknown parameter. It helps to show the flexibility of choosing the initial guess in the frame of HAM.

Using (5.6) and (5.7), define non-linear operator as

$$N^*[\psi(\eta;p)]=\psi'''(\eta)+\frac{1+k}{2}\psi(\eta)\psi''(\eta)-k\left(\psi(\eta)\right)^2 \tag{5.9}$$

The zeroth order equation in HAM process is given by

$$\begin{aligned}&(1-p)L^*[\psi(\eta;p)-f_0(\eta)]=c_0pN^*[\psi(\eta;p)]\\ \text{with B. C.}\quad &\psi(\eta)=0, \quad \psi'(\eta)\,|_{\eta=0}=1 \quad \& \quad \psi'(\eta)\,|_{\eta\to\infty}=0\end{aligned} \tag{5.10}$$

The mth-order deformation equation can be constructed as

$$L^*\left[f_m(\eta)-\chi_m f_{m-1}(\eta)\right]=c_0D_{m-1}\left\{N^*[\psi(\eta;p)]\right\} \tag{5.11}$$

Along with proper $f_0(\eta)$, L^*, and c_0, the homotopy series absolutely converges at $p=1$, and thus

$$f(\eta)=f_0(\eta)+\sum_{m=1}^{\infty}f_m(\eta) \tag{5.12}$$

Here,

$$f_m(\eta)=\frac{1}{m!}\frac{d^m\psi(\eta;p)}{dp^m}\,|_{p=0}$$

$$\text{Subject to B. C.,}\quad f_m(0)=0, \quad f_m'(0)=0, \quad f_m'(\infty)=0$$

The general solution of mth-order deformation equation is

$$f_m(\eta) = f^*_m(\eta) + B_0 + B_1 e^{-\eta} + B_2 e^{\eta} \tag{5.13}$$

Using boundary conditions of deformation equation values of B_0, B_1 and B_2 are

$$B_0 = -\left(f^*_m(0) + f^*_m(\eta)\,|_{p=0}\right), \quad B_1 = f'^*_m(\eta)\,|_{p=0}, \quad B_2 = 0$$

Hence by substituting these values in the solution of the deformation equation,

$$f_m(\eta) = f^*_m(\eta) - \left(f^*_m(0) + f^*_m(\eta)\,|_{p=0}\right) + f'^*_m(\eta)\,|_{p=0}\, e^{-\eta} \tag{5.14}$$

Thus, homotopy series solution of boundary value problems (5.6) and (5.7) is given by

$$f(\eta) = f_0(\eta) + \sum_{m=1}^{\infty} f_m(\eta)$$

$$f(\eta) = \left[\sigma^* + (1 - 2\sigma^*)e^{-\eta} - (1 - \sigma^*)e^{-2\eta}\right]$$

$$+ \sum_{m=1}^{\infty} \left(f^*_m(\eta) - \left(f^*_m(0) + f^*_m(\eta)\,|_{p=0}\right) + f'^*_m(\eta)\,|_{p=0}\, e^{-\eta}\right) \tag{5.15}$$

5.4 RESULT AND DISCUSSION

The residual error is displayed with respect to iteration which tells the information about the number of iterations required for a suitable computational execution process. The respective homotopy approximation contains two unknown parameters: σ^* and the control parameter c_0. The found optimal convergence-control parameter is $c_0^* = -1.2411$ and the optimal parameter $\sigma^* = 3.0435$. Figure 5.1 shows the variation between the square residual and the number of iterations.

Figure 5.1 illustrates that after the 6th iteration, the solution gives desired result for $c_0 = -5/4$ and different (series 1) $k = -1/4$, $\sigma = 3$, (series 2) $k = -1/3$, $\sigma^* = 3$, (series 4) $k = -1/4$, $\sigma^* = 11/4$. Except $k = 0$, (series 3) $\sigma = 11/4$, square residual approaches to zero at 6th iteration. After the 12th iteration, the series converges. For instance here, $c_0 = -\frac{5}{4}$, $\sigma^* = 3$ is considered for computation. Variation of k is shown in Figure 5.2. Series 1, 2, 3, 4, 5, and 6 show the variation between f and η for $k = -1/2, -1/4, -1/5, -1/6, -1/7$

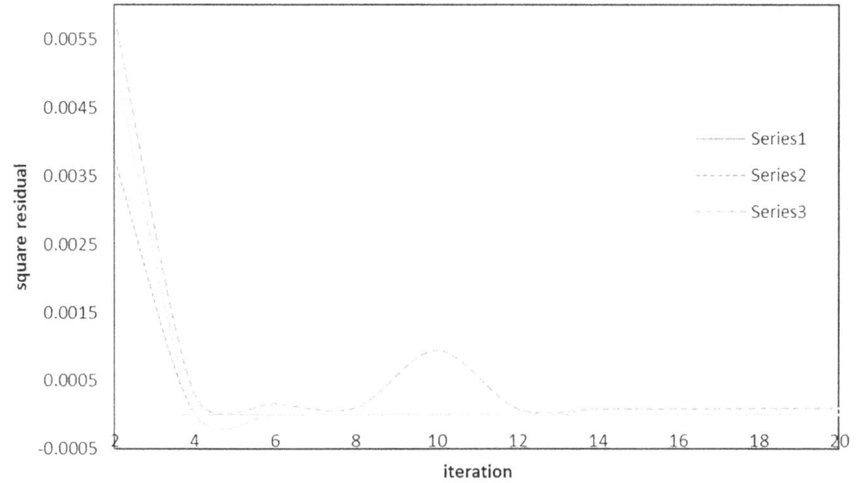

Figure 5.1 Square residual vs iteration.

and 0, respectively. Figure 5.2 demonstrates the variation of solution (5.15) of equations (5.6) and (5.7) for different k with respect to η keeping c_0 = −5/4 and $\sigma^* = 3$. It is found that the value of k must be greater than −1/2. As k increases from −1/2 to 0, the solution is decreasing. A similar result is found in Banks [11]. Figure 5.3 explores the behaviour of the velocity of the fluid showing variation between f' and η for various value k and for constant $c_0 = -5/4$, $\sigma^* = 3$. Series 1, 2, 3, 4, 5 show the variation of $k = -1/2$, −1/5, −1/6, −1/7, 0, respectively.

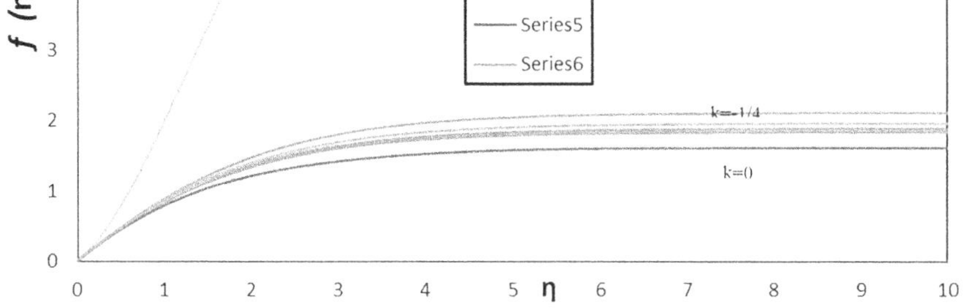

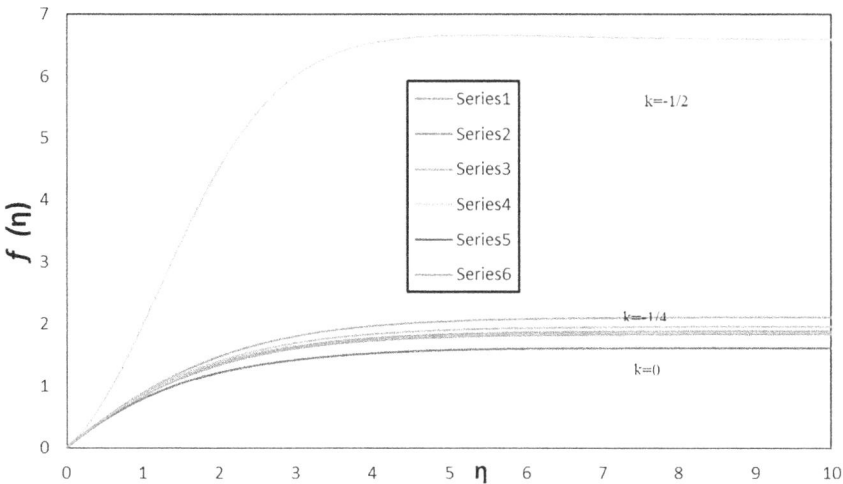

Figure 5.2 vs η for different $k, c_0 = -\frac{5}{4}, \sigma = 3$.

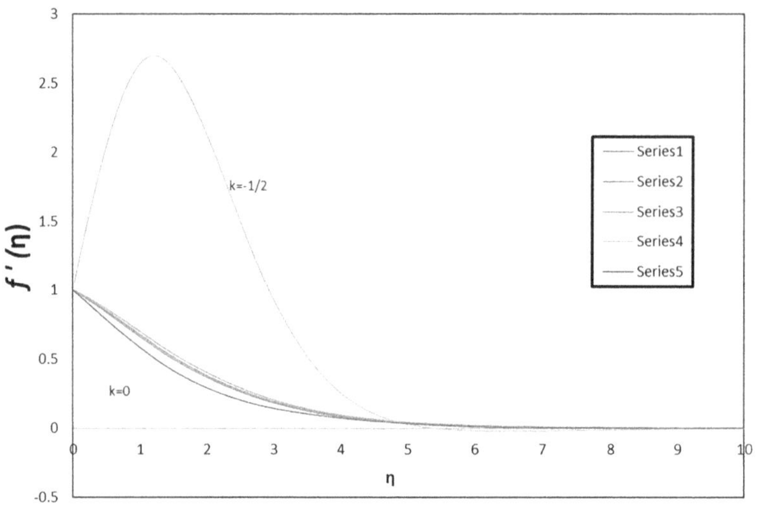

Figure 5.3 vs η for different $k, c_0 = -\frac{5}{4}, \sigma = 3$.

The velocity profile for the fluid flow clearly satisfies the boundary condition (5.7). Again for $k < -1/2$, it is showing the undesired result and the 10th-order homotopy approximation of $f(\eta)$ agrees well with the exact solution [10,11].

Table 5.1 shows the variation of skin coefficient friction, square residual and Central processing unit (CPU) time when $c_0 = -\frac{5}{4}$, $k = -\frac{1}{4}$ and $\sigma^* = 3$. It is observed from the Table 5.1 that the skin coefficient friction $(-f''(0))$ is converges to 0.1620. Further, the square residual is decreasing to zero up to 10 decimal place as increase in order of approximation hence giving the appropriate value for skin friction on the wall. Moreover, it is noticed that the CPU time taken for the calculation increased while the order of approximations increased.

Table 5.2 shows the variation of skin coefficient friction, square residual and CPU time when $c_0 = -5/4$, $k = -\frac{1}{3}$, $\sigma^* = 3$. It is observed from Table 5.2, CPU time increases with the order of approximation which means it takes more time to evaluate a higher order of approximation because it uses recurrence to solve higher order approximations.

Table 5.3 shows the variation of skin coefficient friction, square residual and CPU time when $c_0 = -\frac{5}{4}$, $k = -\frac{1}{4}$, $\sigma^* = \frac{11}{4}$. It is observed from the Table 5.2 that the skin coefficient friction $(-f''(0))$ is converges to 0.1620 and the square residual is decreasing to zero (up to 10 decimal place) at the 18th order of approximation. The same behaviour is observed in Table 5.4.

Table 5.1 Calculation of skin coefficient friction, square residual and CPU time when $c_0 = -\frac{5}{4}$, $k = -\frac{1}{4}$ and $\sigma^* = 3$

Order of approximation (m)	*f″(0)*	*Square residual*	*CPU time (in sec)*
2	−0.09338766431051587	$5.120578688696486 \times 10^{-3}$	1.3750331
4	−0.16220345423058216	$3.950912302849012 \times 10^{-5}$	5.8445509
6	−0.16690872850287694	$1.146109934474865 \times 10^{-5}$	13.2387715
8	−0.1652100410522276	$3.156986921036227 \times 10^{-7}$	24.8795267
10	−0.16391978312196026	$1.441579999425422 \times 10^{-7}$	43.9625375
12	−0.16304400600369115	$3.533936687814996 \times 10^{-8}$	73.4289095
14	−0.1625664965810511	$1.347689294174587 \times 10^{-8}$	117.0931881
16	−0.16228699224728602	$3.274644136487905 \times 10^{-9}$	190.0532815
18	−0.16214969546020697	$9.64484633042840 \times 10^{-10}$	468.6444263
20	−0.16207715192929567	$2.13827346888059 \times 10^{-10}$	904.4101228

Table 5.2 Calculation of skin coefficient friction, square residual and CPU time when $c_0 = -5/4$, $k = -\frac{1}{3}$, $\sigma^* = 3$

Order of approximation (m)	*f″(0)*	*Square residual*	*CPU time (in sec)*
2	0.07005070546737213	$5.2020033226758 \times 10^{-3}$	0.9218826
4	−0.020875178619019025	$3.70921757314193 \times 10^{-5}$	2.9687811
6	−0.014128983638727135	$3.76663246612728 \times 10^{-6}$	7.7733135
8	−0.008862325932251022	$1.77133790422757 \times 10^{-6}$	14.3087992
10	−0.004789754052512101	$6.79084833320169 \times 10^{-7}$	24.2793975
12	−0.0024073075765823817	$2.36195915020101 \times 10^{-7}$	38.4063812
14	−0.0010688715484158768	$7.32943115439545 \times 10^{-8}$	58.6053548
16	−0.00037301567350529643	$2.06579271815528 \times 10^{-8}$	87.4581349
18	−0.00004612579389389109	$5.32136383355773 \times 10^{-9}$	126.5569816
20	0.00008221409158140948	$1.36449086872506 \times 10^{-9}$	178.9276979

5.5 CONCLUSION

The flow induced by a moving flat surface is described by partial differential equations, which are transformed into a single ordinary differential equation using a suitable similarity transformation. The robust homotopy series method is then employed to solve this equation. The study examines the impact of an unknown parameter $\dot{\sigma}^*$ on velocity and skin friction coefficient. Residual error with respect to iteration is also elaborated. The outcomes of the study are as follows:

Table 5.3 Calculation of skin coefficient friction, square residual and CPU time when $c_0 = -\frac{5}{4}, k = -\frac{1}{4}, \sigma^* = \frac{11}{4}$

Order of approximation (m)	*f″(0)*	*Square residual*	*CPU time (in sec)*
2	−0.1089249716864691 8	0.003754108309057125 7	1.5408271
4	−0.16768608977548477	0.00002010660795029711	6.0445518
6	−0.1681395035495125	0.000001344202431788572	13.7189766
8	−0.165571451788681 96	$4.048578107551603\times10^{-7}$	26.1502146
10	−0.163861192208149 24	$1.377951424027182\times10^{-7}$	46.0845858
12	−0.16291130763763012	$3.644158912320677\times10^{-8}$	77.6166253
14	−0.16241707740014147	$9.432714670622303\times10^{-9}$	125.7298804
16	−0.16217640849663145	$2.085748864810874\times10^{-9}$	200.3763935
18	−0.1620693261711835	$4.312551640601 85\times10^{-10}$	559.3550385
20	−0.1620278647083035	$8.71220064983422\times10^{-11}$	972.0297493

Table 5.4 Calculation of skin coefficient friction, square residual when $k = 0$, $c_0 = -\frac{5}{4}$, $\sigma^* = \frac{11}{4}$

Order of approximation (m)	*f″(0)*	*Square residual*
2	−0.3724268353174603	0.005807649988059016
4	−0.4365696305906429	0.0002903060341792165
6	−0.4354530587019302	0.0001689143937700651
8	−0.45136082648835696	0.0001076963911581059 1
10	−0.4403294889708924	0.0000941887185082419 4
12	−0.44329729504982235	0.0000873068775382398 4
14	−0.4473680091225853	0.0000834228152920154 5
16	−0.4402652692931631	0.0000850178703865574 4
18	−0.4448792968032122	0.0000903815582212251 6
20	−0.44579465949068187	0.0000956723306450896 9

- The residual errors at every iteration during the solution process are examined, and it is found that the series generated by HAM converges and stabilizes after the 12th iteration, indicating the accuracy and effectiveness of the method.
- Parameter $\dot{\sigma}^*$ helps in exploring the flexibility of choosing initial approximation in HAM. Due to recurrence, the computational time for higher order approximation is more.

NOMENCLATURE

σ^*	Unknown parameters for flexibility of initial guess
ϑ	Kinematic viscosity
χ	Stream function
A, B	Positive constant
c_0	Convergence-control parameter
$f_0(\eta)$	Initial guess
L^*	Auxiliary linear operator
N^*	Non-linear operator
U	Velocity component in X direction
$U_w(X) = A(X+B)^k$	The speed of flat plate at wall
V	Velocity component in Y direction

REFERENCES

[1] Kumaran, V., & Ramanaiah, G. (1996). A note on the flow over a stretching sheet. *Acta Mechanica*, *116*(1), 229–233.

[2] Sakiadis, B. C. (1961). Boundary-layer behavior on continuous solid surfaces: II. The boundary layer on a continuous flat surface. *AiChE Journal*, *7*(2), 221–225.

[3] Crane, L. J. (1970). Flow past a stretching plate. *Zeitschrift für angewandte Mathematik und Physik ZAMP*, *21*, 645–647.

[4] Liao, S. (2003). *Beyond Perturbation - Introduction to the Homotopy Analysis Method*, Chapman & Hall/CRC Press, Boca Raton.

[5] Vasu, B., Ray, A. K., & Gorla, R. S. (2020). Free convective heat transfer in Jeffrey fluid with suspended nanoparticles and Cattaneo-Christov heat flux. *Proceedings of the Institution of Mechanical Engineers, Part N: Journal of Nanomaterials, Nanoengineering and Nanosystems*, *234*(3–4), 99–114.

[6] Liao, S. (2012). *Homotopy Analysis Method in Nonlinear Differential Equations*. Higher Education Press, Beijing.

[7] Liao, S. (2004). On the homotopy analysis method for nonlinear problems. *Applied Mathematics and Computation*, *147*(2), 499–513.

[8] Ray, A. K., Vasu, B., & Gorla, R. S. R. (2019), Homotopy simulation of non-Newtonian spriggs fluid flow over a flat plate with oscillating motion. *International Journal of Applied Mechanics and Engineering*, *24*(2), 359–385.

[9] Bég, O. A., Vasu, B., Ray, A. K., Bég, T. A., Kadir, A., Leonard, H. J., & Gorla, R. S. (2020). Homotopy simulation of dissipative micropolar flow and heat transfer from a two-dimensional body with heat sink effect: applications in polymer coating. *Chemical and Biochemical Engineering Quarterly*, *34*(4), 257–275.

[10] Liao, S. J., & Pop, I. (2004). Explicit analytic solution for similarity boundary layer equations. *International Journal of Heat and Mass Transfer*, *47*(1), 75–85.

[11] Banks, W. H. H. (1983). Similarity solutions of the boundary-layer equations for a stretching wall. *Journal de Mécanique théorique et appliquée*, *2*(3), 375–392.

Chapter 6

Modeling of electrically conducting fluid flow over a rotating disk

Anupam Bhandari and Akmal Husain
University of Petroleum and Energy Studies

6.1 INTRODUCTION

Colloidal suspensions of magnetic particles are known as ferrofluids [1]. Commercial applications include sealing the spinning shaft, lubrication, and managing wire heat. Recent uses of this smart fluid include cancer treatment and targeting drugs [2]. The main components of ferrofluids are carrier liquids, magnetic nanoparticles, and surfactants. Normally, when a magnetic field exists, the magnetic force has an impact on the flow [3]. Force requires material magnetization to be magnetic. However, in a rotating flow, the fluid is affected by the magnetic torque because the particles and the fluid rotate at different speeds. The Lorentz force is crucial in determining the flow characteristics of ferrohydrodynamic flow if this fluid is electrically conducting. As a result of the distinction in speed between magnetic particles and fluid, the magnetic torque may cause the flow to become more viscous. However, the liquid's viscosity also depends on the temperature of the substance. In this chapter, we will address the role of temperature-based viscosity on velocity and temperature patterns.

Kafoussias and Williams [4] inspected the impact of temperature-based viscosity on the flow of a laminar boundary layer over a vertical semi-infinite plate. They concluded that the role of this viscosity should also be considered in the mathematical calculations; otherwise, some errors can occur in the flow characteristics. Elbashbeshy and Bazid [5] looked into the flow over a continuously moving surface and reported that the variable temperature-based viscosity parameter raises the temperature and decreases the skin resistance to the surface. In a microconvective flow, Kumar and Mahulikar [6] reported that temperature-dependent viscosity qualitatively changes both the speed of the flow and temperature. Nadeem and Akbar [7] represented a peristaltic laminar Jeffrey fluid flow in an uneven vertical tube to study the impact of variable temperature-based viscosity. Hooman and Gurgenc [8] demonstrated the convective flow in a porous channel with variable temperature-based viscosity and focused on the influence of the

 DOI: 10.1201/9781032712079-6

Péclet number and Prandtl number regarding the flow properties. Booker [9] presented the fundamental aspect of variable temperature viscosity. Devi and Prakash [10] found that at high temperatures, variable viscosity and conductivity have an important role in hydromagnetic fluid flow resulting from a stretched sheet.

Mille et al. [11] analyzed the stability analysis of the movement of the boundary layer over a spinning surface with variable temperature-based viscosity. Khan et al. [12] examined the Newtonian fluid's physical characteristics along with varying conductivity and viscosity on the flow near a rotating surface. Iqra and Mustafa [13] scrutinized the different viscosity models for decelerating rotating flow problems with variable physical properties. Khan et al. [14] considered that viscosity is oppositely related to temperature and measured the thermo-physical behavior of the fluid near a rotating disk. For the convective flow problem around a disk, Maleque [15] explored how temperature affects a situation and depth-dependence viscosity on the patterns of speed and temperature along with skin friction coefficient and local heat transfer.

In this paper, we investigate the movement of an incompressible fluid that conducts electric charge over a spinning disk in a heightened state of the Lorentz and Kelvin forces. This study specifically examines how the Lorentz force, temperature-based variable viscosity, and variable conductivity affect the heat and momentum boundary layers. The Navier–Stokes equation in cylindrical coordinates is used in this instance, and the ordinary differential equations are created through similarity transformation from the governing equations. The COMSOL Multiphysics software is then used to numerically solve these equations utilizing the finite element technique. The findings for patterns of velocity and temperature are obtained, and a detailed discussion of the impact of the temperature-based viscosity parameter, variable conductivity parameter, and Hartman number on the flow and thermal properties is provided.

6.2 THEORETICAL MODEL

The flow under consideration is schematically shown in Figure 6.1. With the angular velocity ω, the disk rotates consistently. The expressions u, v, and w represent the radial, azimuthal, and axial velocity along r, φ, and z directions, respectively. The surface temperature is T_w and away from the surface is T_c. The expression B denotes the magnetic induction. The following equations determine the current flow:

$$\frac{\partial u}{\partial r}+\frac{u}{r}+\frac{\partial w}{\partial z}=0 \tag{6.1}$$

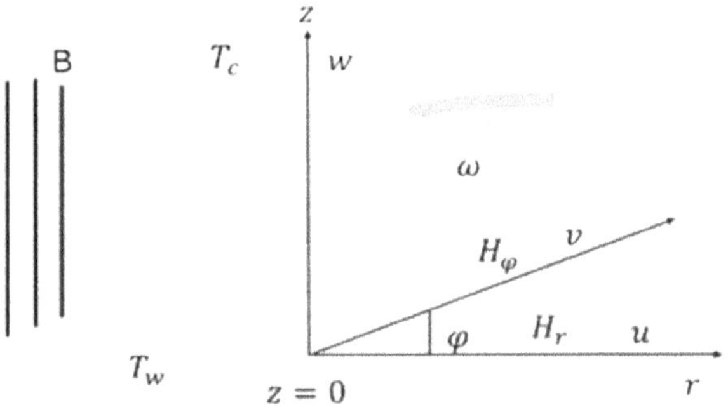

Figure 6.1 An infinitely rotating disk creates a flow arrangement of electrically conducting magnetic fluid.

$$\rho\left[u\frac{\partial u}{\partial r}+w\frac{\partial u}{\partial z}-\frac{v^2}{r}\right]=-\frac{\partial p}{\partial r}+\left\{\frac{\partial}{\partial r}\left(\mu_1\frac{\partial u}{\partial r}\right)+\frac{\partial}{\partial r}\left(\mu_1\frac{u}{r}\right)+\frac{\partial}{\partial z}\left(\mu_1\frac{\partial u}{\partial z}\right)\right\}$$
$$-\mu_0 M\frac{\partial H}{\partial r}-\sigma B^2 u \tag{6.2}$$

$$\rho\left[u\frac{\partial v}{\partial r}+w\frac{\partial v}{\partial z}+\frac{uv}{r}\right]=\left\{\frac{\partial}{\partial r}\left(\mu_1\frac{\partial v}{\partial r}\right)+\frac{\partial}{\partial r}\left(\mu_1\frac{v}{r}\right)+\frac{\partial}{\partial z}\left(\mu_1\frac{\partial v}{\partial z}\right)\right\}$$
$$-\mu_0\frac{M}{r}\frac{\partial H}{\partial \varphi}-\sigma B^2 v \tag{6.3}$$

$$\rho c_p\left(u\frac{\partial T}{\partial r}+w\frac{\partial T}{\partial z}\right)=\frac{k_1}{r}\frac{\partial T}{\partial r}+\frac{\partial}{\partial r}\left(k_1\frac{\partial T}{\partial r}\right)+\frac{\partial}{\partial z}\left(k_1\frac{\partial T}{\partial z}\right)$$
$$+\mu_1\left\{\left(\frac{\partial u}{\partial z}\right)^2+\left(\frac{\partial v}{\partial z}\right)^2\right\}+\sigma B^2\left(u^2+v^2\right) \tag{6.4}$$

In the above equations, ξ exhibits the Langevin parameter, φ_1 indicates the quantity of concentration, p represents the fluid pressure, μ_1 indicate the temperature-dependent viscosity, μ_0 indicate the open space's permeability, M indicates the magnetization, σ indicates the electric conductivity, ρ indicates the density of the substance, c_p indicate the heat capacity at constant pressure, k_1 stands for variable thermal efficiency and T indicates the temperature.

The initial and boundary conditions considering no slip effects are as follows:

$$z=0;\ \begin{cases} u=0 \\ v=r\omega \\ w=0 \\ T=T_w \end{cases} \quad \text{and} \quad z\to\infty;\ \begin{cases} u=0 \\ v=0 \\ T=T_c \end{cases} \tag{6.5}$$

The temperature-dependent viscosity [16] can be calculated as

$$\mu_1=\frac{\mu}{1+\alpha(T-T_c)}, \tag{6.6}$$

where μ denotes the fluid viscosity and α represents the thermal expansion factor.

The variable conductivity [17] is defined as

$$k_1=k\left(1+\epsilon\frac{T-T_c}{T_w-T_c}\right), \tag{6.7}$$

where k is a symbol for heat conductivity and ϵ is the variable conductivity parameter.

The magnetic field [18]

$$H=\frac{\xi_0}{2\pi}\frac{1}{r^2}, \tag{6.8}$$

where ξ_0 shows how strong the magnetic field is.

Linear relation between magnetization and temperature [19,20]:

$$M=K(T_c-T), \tag{6.9}$$

where the parameter K denotes the paramagnetism coefficient.

We use the following similarity transformation [21–23]:

$$\begin{gathered}\eta=\sqrt{\frac{\omega}{v}}z,\quad u=r\omega E(\eta),\quad v=r\omega F(\eta),\quad w=\sqrt{\omega v}\ G(\eta),\\ p=\rho\omega v P(\eta),\quad T=T_c+(T_w-T_c)\theta(\eta),\end{gathered} \tag{6.10}$$

where η indicates the dimensionless distance; E, F, and G indicate the dimensionally unrestricted radial, azimuthal, and axial velocity, respectively; and θ indicates the dimensionless temperature.

Using the above similarity transformation, we obtain the following nonlinear differential equations:

$$\frac{dG}{d\eta}+2E=0 \tag{6.11}$$

$$(1-b\theta)\frac{d^2E}{d\eta^2}-b\frac{dE}{d\eta}\frac{d\theta}{d\eta}-G\frac{dE}{d\eta}-E^2+F^2-M_1E-\beta\theta=0 \tag{6.12}$$

$$(1-b\theta)\frac{d^2F}{d\eta^2}-b\frac{dF}{d\eta}\frac{d\theta}{d\eta}-G\frac{dF}{d\eta}-2EF-M_1F=0 \tag{6.13}$$

$$(1+\epsilon\theta)\frac{d^2\theta}{d\eta^2}+\epsilon\left(\frac{d\theta}{d\eta}\right)^2-\Pr G\frac{d\theta}{d\eta}+Ec\left[\left(\frac{dE}{d\eta}\right)^2+\left(\frac{dF}{d\eta}\right)^2\right]$$
$$+Ec\ M_1\left(E^2+F^2\right)=0 \tag{6.14}$$

$$E(0)=0,\quad F(0)=1,\quad G(0)=0,\quad \theta(0)=1;\quad E(\infty)=0,\quad F(\infty)=0,\quad \theta(\infty)=0 \tag{6.15}$$

The non-dimensional parameters are defined as follows:

$$b=\alpha(T_w-T_c),\quad \beta=\frac{\mu_0\xi_0K(T_c-T_w)}{2\pi\rho r^4\omega^2},\quad \Pr=\frac{\mu c_p}{k},\quad Ec=\frac{r^2\omega^2}{(T_w-T_c)c_p},$$

$$M_1=\frac{\sigma B^2}{\rho\omega}$$

6.3 NUMERICAL SOLUTION AND RESULTS

We use the finite element method for solving nonlinear differential equations numerically. For the solution, we use the weak form of the similarity differential equations. For a typical element, we consider the quadratic Lagrange's interpolation functions. The element size is considered 0.001. For the iteration process, we use the Newton method in COMSOL Multiphysics, and a maximum of 25 iterations are considered. Considering the parameter $b=0$, $M_1=0$, $\beta=0$, and $\epsilon=0$, the current problem can be converted to the original Von-Karman problem of the swirling flow. In this chapter, we have measured the influence of the variable temperature-based viscosity, variable conductivity, and Lorentz force in relation to the temperature and velocity patterns.

In the present case, the fluid layer closest to the disk is pulled along and pushed outward by centrifugal force. Following that, fresh fluid particles are continually drawn onto the disk in an axial direction before being centrifugally expelled once more. Consequently, this is a fully three-dimensional flow that serves as a pump. Despite the fact that the computation is only technically relevant to a disk that is indefinitely expanded, the results will now be applied to a disk with a defined radius. Figure 6.2 represents the impact of the temperature-dependent viscosity variable affecting boundary layer flow. The radial, azimuthal, and axial velocities as well as temperature variation are shown in the result. The temperature-based viscosity variable (b) reduces the flow velocity and enhances the temperature in the boundary layer. The negative range of axial velocity stands for which way the flow is going toward the disk.

Figure 6.3 demonstrates the role of variable conductivity parameters based on velocity components and temperature of the considered boundary layer flow. The variable conductivity parameter (ϵ) also lowers the speed and makes the temperature higher. The variable conductivity favors the shear thickening of the fluid; therefore, it has similar characteristics as the temperature-based viscosity. However, this parameter shrinks the velocity less than the temperature-dependent viscosity parameter. Figure 6.4 represents how the magnetic parameter (M_1) affects the momentum and thermal boundary layer. This effect occurs due to the presence of the conductive nature of the fluid. This parameter reduces the velocity in the flow since

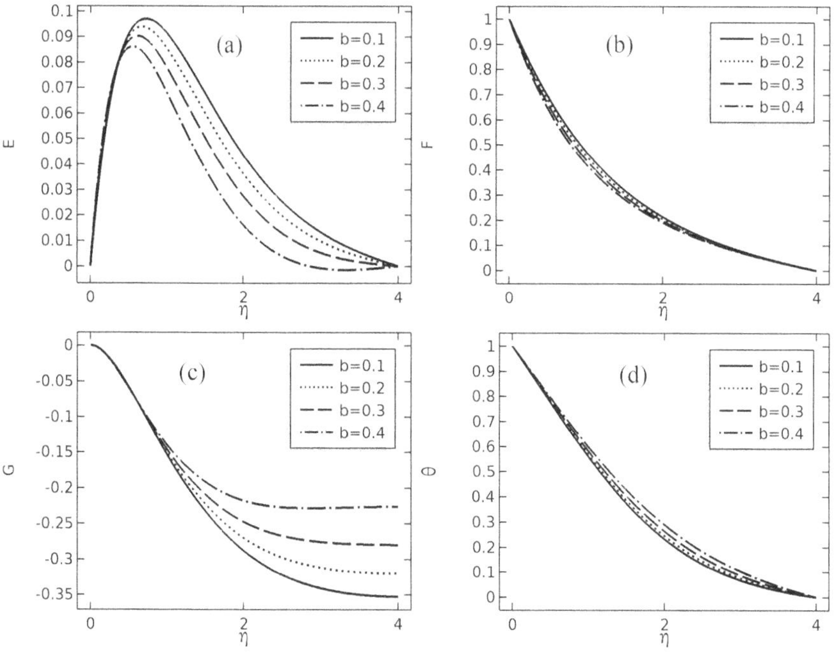

Figure 6.2 Effect of a variable temperature-viscosity parameter on the temperature and velocity curves.

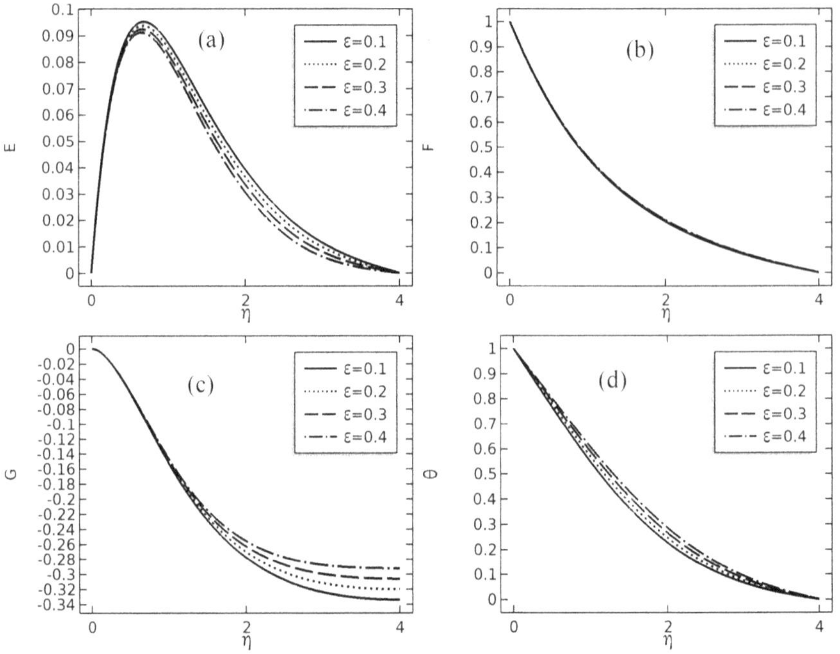

Figure 6.3 Curves of velocity and temperature are affected by a changeable conductivity parameter.

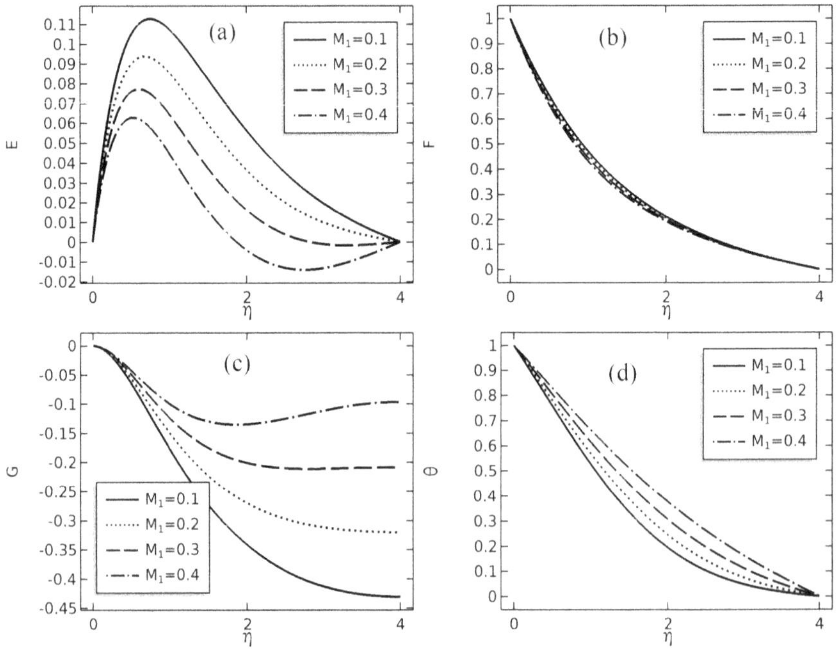

Figure 6.4 Influence of Lorentz force on the velocity and temperature curves.

this force opposes the flow and reduction in the velocity favors the thermal boundary layer in the flow. Therefore, the temperature is reduced in the flow.

6.4 CONCLUDING REMARK

In this chapter, flowing over a rotating surface in a laminar boundary layer is numerically investigated to measure the effects of fluctuating viscosity, variable conductivity, and Lorentz force on the velocity and temperature profiles. Our study shows that the presence of viscosity that changes with temperature and in the flow, changing conductivity decreases the momentum boundary layer and intensifies the heat boundary layer. If the fluid is conducting, then the Lorentz force significantly affects the flow rate and temperature dispersion. However, for non-conducting fluid, this force will not affect this boundary layer flow.

REFERENCES

[1] A. Bhandari and K. P. S. Parmar, "Influence of magnetic dipole on ferrohydrodynamic thin film flow over an inclined spinning surface," *Phys. Fluids*, vol. 35, no. 2, p. 22008, Jan. 2023, doi:10.1063/5.0138600.

[2] A. Bhandari, "Theoretical development in the viscosity of ferrofluid," *J. Tribol.*, vol. 145, no. 5, 2023, doi:10.1115/1.4056626.

[3] P. Ram and A. Bhandari, "Negative viscosity effects on ferrofluid flow due to a rotating disk," *Int. J. Appl. Electromagn. Mech.*, vol. 41, no. 4, pp. 467–478, 2013, doi:10.3233/JAE-121637.

[4] N. G. Kafoussias and E. W. Williams, "The effect of temperature-dependent viscosity on free-forced convective laminar boundary layer flow past a vertical isothermal flat plate," *Acta Mech.*, vol. 110, no. 1, pp. 123–137, 1995, doi:10.1007/BF01215420.

[5] E. M. A. Elbashbeshy and M. A. A. Bazid, "The effect of temperature-dependent viscosity on heat transfer over a continuous moving surface," *J. Phys. D Appl. Phys.*, vol. 33, no. 21, p. 2716, 2000, doi:10.1088/0022-3727/33/21/309.

[6] R. Kumar and S. P. Mahulikar, "Effect of temperature-dependent viscosity variation on fully developed laminar microconvective flow," *Int. J. Therm. Sci.*, vol. 98, pp. 179–191, 2015, doi:10.1016/j.ijthermalsci.2015.07.011.

[7] S. Nadeem and N. S. Akbar, "Effects of temperature dependent viscosity on peristaltic flow of a Jeffrey-six constant fluid in a non-uniform vertical tube," *Commun. Nonlinear Sci. Numer. Simul.*, vol. 15, no. 12, pp. 3950–3964, 2010, doi:10.1016/j.cnsns.2010.01.019.

[8] K. Hooman and H. Gurgenci, "Effects of temperature-dependent viscosity on forced convection inside a porous medium," *Transp. Porous Media*, vol. 75, no. 2, pp. 249–267, 2008, doi:10.1007/s11242-008-9220-1.

[9] R. W. Griffiths, "Thermals in extremely viscous fluids, including the effects of temperature-dependent viscosity," *J. Fluid Mech.*, vol. 166, pp. 115–138, 1986, doi:10.1017/S002211208600006X.
[10] S. P. Anjali Devi and M. Prakash, "Temperature dependent viscosity and thermal conductivity effects on hydromagnetic flow over a slendering stretching sheet," *J. Niger. Math. Soc.*, vol. 34, no. 3, pp. 318–330, 2015, doi:10.1016/j.jnnms.2015.07.002.
[11] R. Miller, P. T. Griffiths, Z. Hussain, and S. J. Garrett, "On the stability of a heated rotating-disk boundary layer in a temperature-dependent viscosity fluid," *Phys. Fluids*, vol. 32, no. 2, p. 024105, 2020, doi:10.1063/1.5129220.
[12] M. Khan, T. Salahuddin, and S. O. Stephen, "Variable thermal conductivity and diffusivity of liquids and gases near a rotating disk with temperature dependent viscosity," *J. Mol. Liq.*, vol. 333, p. 115749, 2021, doi:10.1016/j.molliq.2021.115749.
[13] I. Ejaz and M. Mustafa, "A comparative study of different viscosity models for unsteady flow over a decelerating rotating disk with variable physical properties," *Int. Commun. Heat Mass Transf.*, vol. 135, p. 106155, 2022, doi:10.1016/j.icheatmasstransfer.2022.106155.
[14] M. Khan, T. Salahuddin, and S. O. Stephen, "Thermo-physical characteristics of liquids and gases near a rotating disk," *Chaos Solitons Fractals*, vol. 141, p. 110304, 2020, doi:10.1016/j.chaos.2020.110304.
[15] K. A. Maleque, "Effects of combined temperature- and depth-dependent viscosity and hall current on an unsteady MHD laminar convective flow due to a rotating disk," *Chem. Eng. Commun.*, vol. 197, no. 4, pp. 506–521, 2010, doi:10.1080/00986440903288492.
[16] P. G. Siddheshwar, R. K. Vanishree, and A. C. Melson, "Study of heat transport in Bénard-Darcy convection with g-jitter and thermo-mechanical anisotropy in variable viscosity liquids," *Transp. Porous Media*, vol. 92, no. 2, pp. 277–288, Nov. 2012, doi:10.1007/s11242-011-9901-z.
[17] R. Cortell, "Heat transfer in a fluid through a porous medium over a permeable stretching surface with thermal radiation and variable thermal conductivity," *Can. J. Chem. Eng.*, vol. 90, no. 5, pp. 1347–1355, Oct. 2012, doi:10.1002/cjce.20639.
[18] J. L. Neuringer, "Some viscous flows of a saturated ferro-fluid under the combined influence of thermal and magnetic field gradients," *Int. J. Non. Linear. Mech.*, vol. 1, no. 2, pp. 123–137, Oct. 1966, doi:10.1016/0020-7462(66)90025-4.
[19] R. E. Rosensweig, Ferrohydrodynamics, Cambridge University Press, New York, 1985.
[20] A. Bhandari, "Unsteady flow and heat transfer of the ferrofluid between two shrinking disks under the influence of magnetic field," *Pramana*, vol. 95, no. 2, pp. 1–12, May 2021, doi:10.1007/S12043-021-02107-Y.
[21] M. Turkyilmazoglua, "Flow and heat simultaneously induced by two stretchable rotating disks," *Phys. Fluids*, vol. 28, no. 4, p. 043601, Apr. 2016, doi:10.1063/1.4945651.

[22] S. Xinhui, Z. Liancun, Z. Xinxin, and S. Xinyi, "Homotopy analysis method for the asymmetric laminar flow and heat transfer of viscous fluid between contracting rotating disks," *Appl. Math. Model.*, vol. 36, no. 4, pp. 1806–1820, 2012, doi:10.1016/j.apm.2011.09.010.
[23] A. Bhandari, "Entropy generation and heat transfer analysis for ferrofluid flow between two rotating disks with variable conductivity," *Proc. Inst. Mech. Eng. Part C J. Mech. Eng. Sci.*, vol. 235, no. 21, pp. 5877–5891, 2021, doi:10.1177/0954406221991184.

Chapter 7

Unsteady radiative MHD flow over a porous stretching plate

Ankur Kumar Sarma
Cotton University

7.1 INTRODUCTION

Magnetohydrodynamics (MHD) is the scientific study of the interactions between electromagnetic forces and fluid mechanics. Turbulence dampening, modification, and even suppression frequently take place in a variety of flows. To ensure that the product's composition is consistent, a magnetic field, for example, can be employed to alter the flow patterns that happen spontaneously during the formation of single crystals of semiconductors. This seems to be the MHD application area that has the most potential right now, because it has uses in numerous branches of science and technology, including electronics, chemical engineering, and astrophysics. Many scientists are closely examining unsteady MHD laminar flow with heat and mass transfer along a chemically reacting stretched sheet moving with changing velocity in a porous medium.

A flow field produced by a continuous surface flowing at a constant speed was developed by Sakiadis (1961b). Then he investigated the behaviour of the movement of fluid on a continuous flat surface (Sakiadis, 1961a). Crane (1970) examined the flow and heat transmission of a viscous fluid's boundary layer that conducts electricity over a stretching sheet. Equations of boundary layer, for the flow caused only by a stretched surface, were presented and studied by Banks (1983). The impact of transverse magnetic field, radiation is examined by numerous researchers since many natural occurrences and technical challenges are worth exposing to MHD analysis. Then Wang (1984) looked at the 3D flow produced by an extending flat surface. The Navier–Stokes equations' exact similarity solution was discovered. The solution depicted the fluid motion in three dimensions brought on by stretching a flat surface. In the occurrence of a transverse magnetic field, viscoelastic fluid flow across a stretching sheet was researched by Andersson (1992). By establishing a precise analytical solution to the nonlinear boundary layer equation that governs, it has been demonstrated that viscoelasticity has the same effect on flow as an external magnetic field. Elbashbeshy (1998) examined the suction's impact and injection on heat transfer via stretching surface with varied and uniform surface heat flux. Siri et al. (2018) also researched heat transfer over a steady stretching surface

DOI: 10.1201/9781032712079-7

in the occurrence of suction. An investigation was made by Andersson et al. (2000) on the heat transport in a liquid film moved by the horizontal sheet. Raptis et al. (2004), Ghaly (2002), Ishak et al. (2011) and several other researchers evaluated thermal radiation's effects on MHD flow problems via stretching sheet. The effects of axisymmetric stretching on boundary layer flows were investigated by Ariel et al. (2006) using Homotopy perturbation method. It was investigated how an incompressible fluid flows in an unstable boundary layer via a stretching surface with the occurrence of a heat source (Elbashbeshy and Aldawody, 2010). Due to the time dependency of the heat flux and expanding velocity associated with the surface, the flow and temperature fields became unstable. Ahmad looked into the heat transfer and boundary layer movement past an extended plate with changing thermal conductivity (Ahmad et al., 2010). Ishak (2010) also investigated the unsteady MHD laminar flow via stretched plate together with heat transfer. Jhankal et al. (2017) extended Ishak's work by considering the stretching plate in the presence of a porous medium. Rosca et al. (2015) researched the issue of unstable viscous flow across a curved surface. Choudhary et al. (2015) conducted a theoretical investigation to characterise a two-dimensional (2D) unsteady flow across a stretched permeable surface of a viscous incompressible electrically conducting fluid while taking into consideration a constant intensity transverse magnetic field. Alarifi et al. (2019) studied the source effect and MHD flow across a vertically extending sheet. Recently, Megahed et al. (2021) examined the MHD fluid movement caused by a stretching sheet that is not steady in the existence of a porous media, thermal radiation, and fluctuating heat flux. Reddy et al. (2023) investigated how a porous medium's presence affected the way MHD heat transfer fluid flowed through a stretching cylinder. Shah et al. (2023) researched the effect of thermal radiation on convective heat transport in Carreau fluid.

This work is a generalisation of the work done by Ishak (2010). We have considered the issue of unsteady, 2D and laminar flow in the occurrence of transverse magnetic field and radiation over a porous stretching plate. The newly introduced radiation parameter causes considerable fluctuations in the temperature of the fluid. The numerical values of τ and Nu are calculated for distinct values of unsteady parameter $\mathcal{A}$ and proclaimed in tables. The nonlinear controlling PDEs are turned to a collection of nonlinear ODEs using the similarity transformation, which are later on computed with the help of MATLAB bvp4c technique.

7.2 MATHEMATICAL FORMULATION

We examine the flow of a porous stretched plate across a 2D MHD unsteady, laminar boundary layer flow when thermal radiation is present. It is an

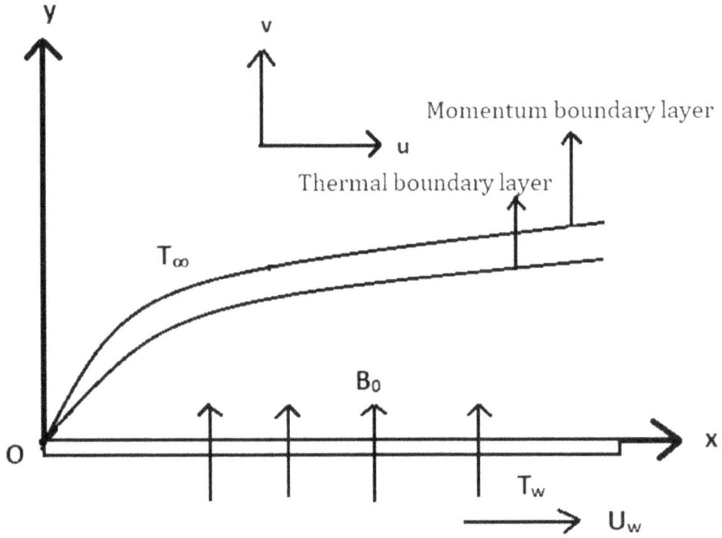

Figure 7.1 Graphical representation of the problem.

electrically conducting, incompressible viscous fluid. Hall effect and polarisation of charges are thought to be caused by the external fluid, which is neglected. The transverse magnetic field B_0 is applied externally perpendicular to x-axis in a positive direction of y-axis (Figure 7.1). The y-axis is normal to the stretching plate, while x-axis is parallel to the stretching plate. The surface is extended with velocity $U_w = \frac{ax}{1-ct}$ along x-axis, maintaining origin fixed.

Under these conditions and Boussinesq's approximation, the governing continuity, momentum and energy equations (Ishak, 2010) are

$$\frac{\partial u}{\partial x}+\frac{\partial v}{\partial y}=0 \tag{7.1}$$

$$\frac{\partial u}{\partial t}+u\frac{\partial u}{\partial x}+v\frac{\partial u}{\partial y}=\nu\frac{\partial^2 u}{\partial y^2}-\frac{\sigma_e B_0^2 u}{\rho}-\frac{\nu u}{k_0} \tag{7.2}$$

$$\frac{\partial T}{\partial t}+u\frac{\partial T}{\partial x}+v\frac{\partial T}{\partial y}=\alpha\frac{\partial^2 T}{\partial y^2}+\frac{\sigma_e B_0^2 u^2}{\rho C_p}-\frac{1}{\rho C_p}\frac{\partial q_r}{\partial y} \tag{7.3}$$

by virtue of boundary conditions (Ishak, 2010) given by

$$\begin{aligned} &y=0: \quad u=U_w, \quad v=0, \quad T=T_w, \quad \text{and} \\ &y\to\infty: \quad u\to 0, \quad T\to T_\infty \end{aligned} \tag{7.4}$$

q_r (Ishak et al., 2011) is described as follows:

$$q_r = -\frac{4\alpha_1}{3k_1}\frac{\partial T^4}{\partial y} \tag{7.5}$$

where α_1 indicates the Stefan-Boltzmann constant and k_1 represents the Rosseland mean absorption coefficient.

We presume that distinction in internal flow temperature is suitably minimal, T^4 can be shown by Taylor series about T_∞ while ignoring the terms with higher order

$$T^4 = 4T_\infty^3 T - 3T_\infty^4 \tag{7.6}$$

We consider the extending velocity $U_w(x,t)$ and surface temperature $T_w(x,t)$ are as follows:

$$U_w = \frac{ax}{1-ct}, \quad T_w = T_\infty + \frac{bx}{1-ct} \tag{7.7}$$

Continuity equation (7.1) is fulfilled by using a stream function ψ such that $u = \frac{\partial \psi}{\partial y}$ and $v = -\frac{\partial \psi}{\partial x}$.

The given nonlinear PDEs (7.2) and (7.3) are transformed into a set of nonlinear ODEs using the following similarity variables and nondimensional quantities given as follows:

$$\eta = \left(\frac{a}{v(1-ct)}\right)^{\frac{1}{2}} y, \quad \psi = \left(\frac{avx^2}{(1-ct)}\right)^{\frac{1}{2}} f(\eta), \quad \theta(\eta) = \frac{T - T_\infty}{T_w - T_\infty}, \quad A = \frac{c}{a},$$

$$M = \frac{\sigma_e B_0^2(1-ct)}{\rho a}, \quad S_p = \frac{v(1-ct)}{k_0 a}, \quad Ec = \frac{a^2 x}{bC_p(1-ct)}, \quad \alpha = \frac{\kappa}{\rho C_p},$$

$$N = \frac{\kappa k_1}{4\sigma_1 T_\infty}, \quad \lambda = \frac{4+3N}{3N}, \quad Pr = \frac{\mu C_p}{\kappa}$$

The transformed nonlinear ODEs are

$$f''' + ff'' - f'^2 - Mf' - S_p f' - A\left(f' + \frac{1}{2}\eta f''\right) = 0 \tag{7.8}$$

$$\frac{\lambda}{Pr}\theta'' + MEcf'^2 + f\theta' - \theta f' - A\left(\theta + \frac{1}{2}\eta\theta'\right) = 0 \tag{7.9}$$

Moreover, the transformed initial and boundary conditions are

$$f(0)=0,\quad f^{'}(0)=1,\quad \theta(0)=1,\quad \text{and}$$
$$f^{'}(\infty)\rightarrow 0,\quad \theta(\infty)\rightarrow 0 \tag{7.10}$$

7.3 METHOD OF SOLUTION

MATLAB solution bvp4c is used to evaluate the system of linear or nonlinear boundary value problems. Unlike the shooting approach, this method is algorithm-based. It can accurately determine the estimated value of $y(x)$ for each x in $[a, b]$ while taking the boundary conditions into account at each stage. By employing this technique, the boundary conditions at infinity are changed for those at a place where the current issue can be logically resolved. To apply finite difference-based solver bvp4c, equations (7.8)–(7.10) are transformed, respectively, as follows:

$$f=y_1,\quad f^{'}=y_1^{'}=y_2,\quad f^{''}=y_2^{'}=y_3,\quad \theta=y_4,\quad \theta^{'}=y_4^{'}=y_5 \tag{7.11}$$

$$y_3^{'}=-y_1y_3+y_2^2+My_2+S_py_2+\mathcal{A}\left(y_2+\frac{1}{2}\eta y_3\right) \tag{7.12}$$

$$y_5^{'}=\frac{Pr}{\lambda}\left[-MEcy_2^2-y_1y_5+y_4y_2+\mathcal{A}\left(y_4+\frac{1}{2}\eta y_5\right)\right] \tag{7.13}$$

Moreover, the initial and boundary conditions (7.10) are transformed as follows:

$$y_1(0)=0,\quad y_2(0)=1,\quad y_4(0)=1 \tag{7.14}$$

$$y_2(\infty)=0,\quad y_4(\infty)=0 \tag{7.15}$$

The above-transformed results are used by the MATLAB solver bvp4c to perform the numerical computation.

7.4 RESULTS AND DISCUSSION

The problem is numerically solved under the above considerations, and behaviours of velocity and temperature are observed for different parameters, including unsteadiness parameter ($\mathcal{A}$), radiation parameter (N), Eckert number (Ec), Hartmann number (M), Prandtl number (Pr), and Porosity parameter (S_P), are shown in graphs in Figures 7.2–7.9.

While the other parameters are fixed, Figures 7.2–7.4 demonstrate the fluctuation relating to velocity profile with regard to distinct values of M, $\mathcal{A}$, and S_P, respectively. As observed, velocity profile drops as M and $\mathcal{A}$ increases, and it steadily decreases as S_P increases. In fact, it is perceived that velocity declines as surface distance escalates and asymptotically meets boundary condition at ∞. Moreover, the τ's magnitude $|-f''(0)|$ rises as the unsteady parameter ($\mathcal{A}$) increases which is shown in Table 7.1. The negative value of $f''(0)$ physically results in a drag force being applied to the fluid by the solid surface. A fluid's flow is always opposed by the drag force. The notion of pure Darcy flow is supported by the drop in velocity profile with an increase in S_P.

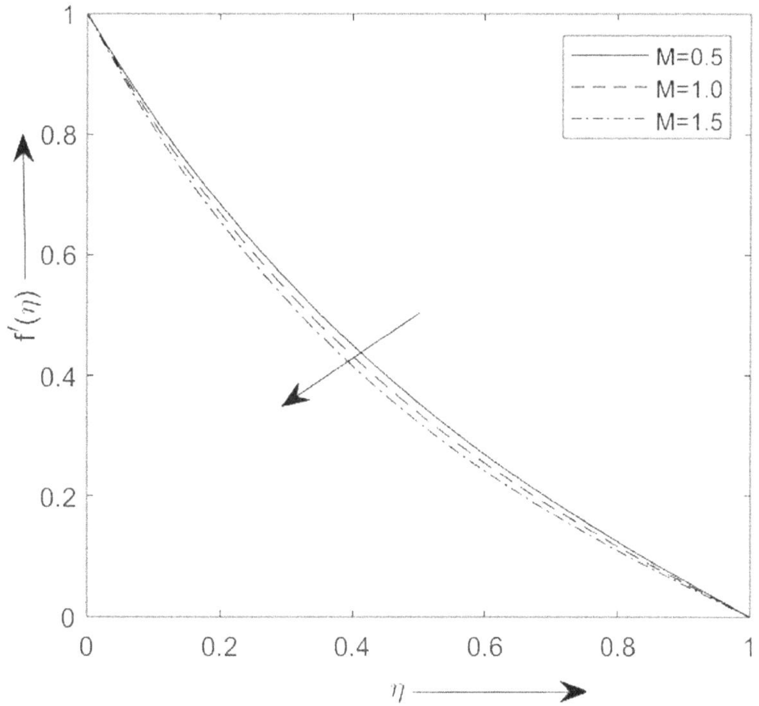

Figure 7.2 V.P (Velocity profile) with the alteration of M when $\mathcal{A} = 1$, $Pr = 1.0$, $S_P = 0.4$, $Ec = 0.05$, $N = 1$.

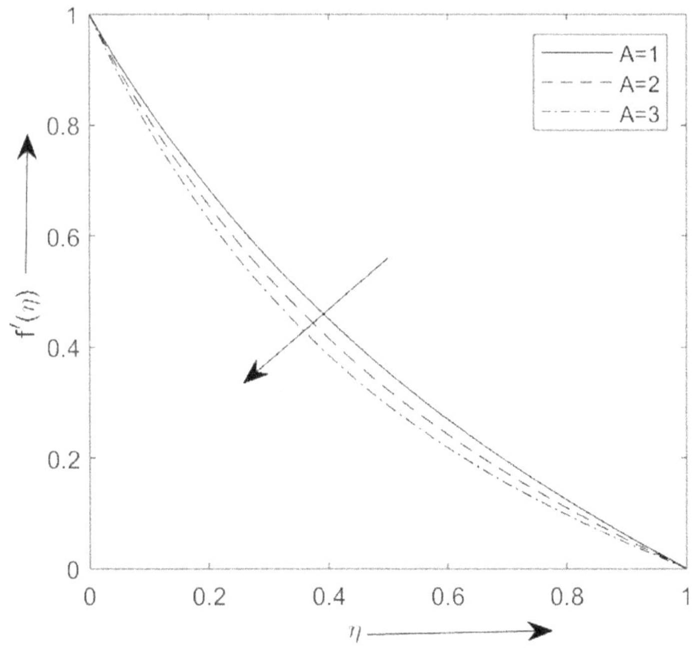

Figure 7.3 V.P with the alteration of $\mathcal{A}$ when $M = 0.5$, $Pr = 1.0$, $S_P = 0.4$, $Ec = 0.05$, $N = 1$.

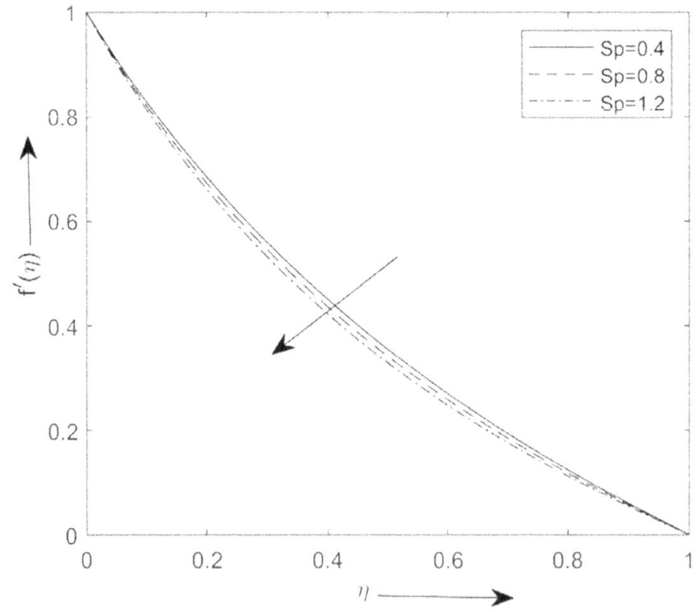

Figure 7.4 V.P with the alteration of S_P when $M = 0.5$, $Pr = 1.0$, $\mathcal{A} = 1$, $Ec = 0.05$, $N = 1$.

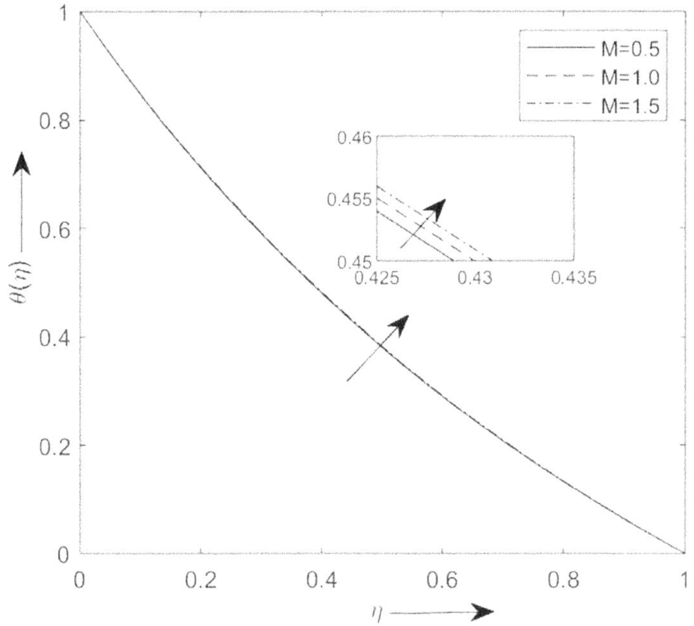

Figure 7.5 T.P (Temperature profile) with the alteration of M when $\mathcal{A} = 1$, $Pr = 1.0$, $S_P = 0.4$, $Ec = 0.05$, $N = 1$.

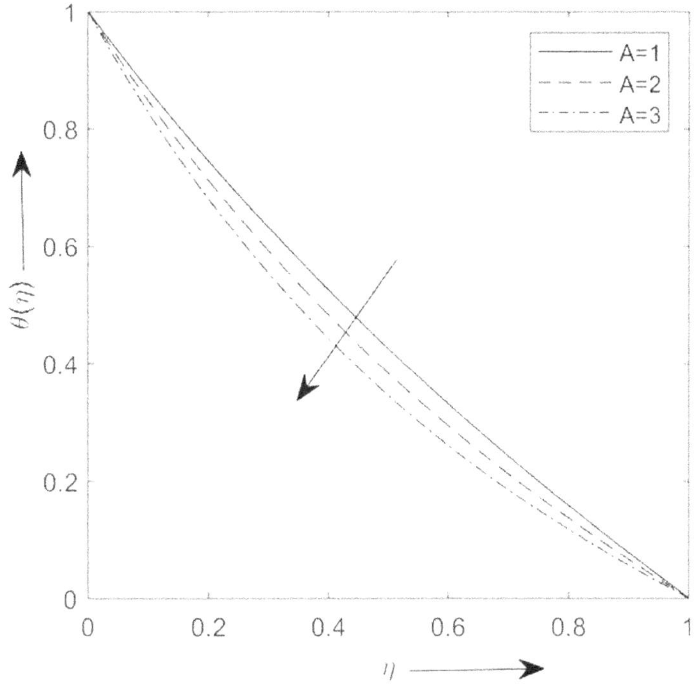

Figure 7.6 T.P with the alteration of $\mathcal{A}$ when $M = 0.5$, $Pr = 1.0$, $S_P = 0.4$, $Ec = 0.05$, $N = 1$.

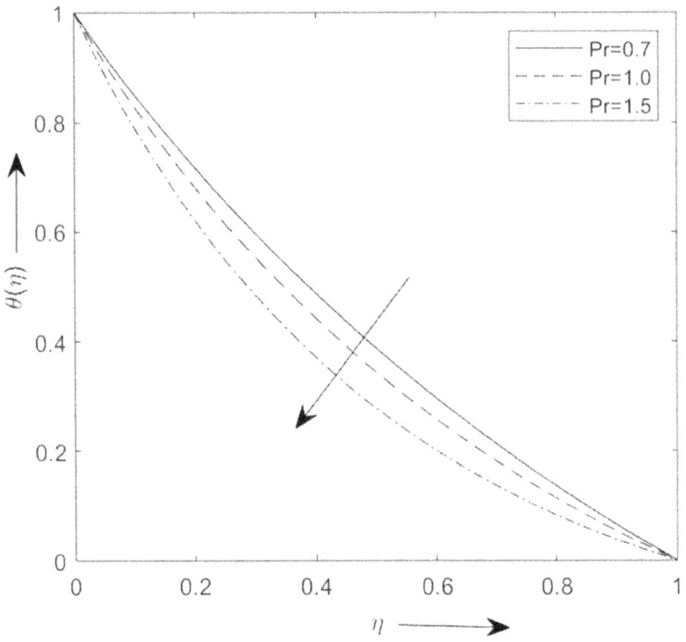

Figure 7.7 T.P with the alteration of *Pr* when $\mathcal{A} = 1$, $M = 0.5$, $S_P = 0.4$, $Ec = 0.05$, $N = 1$.

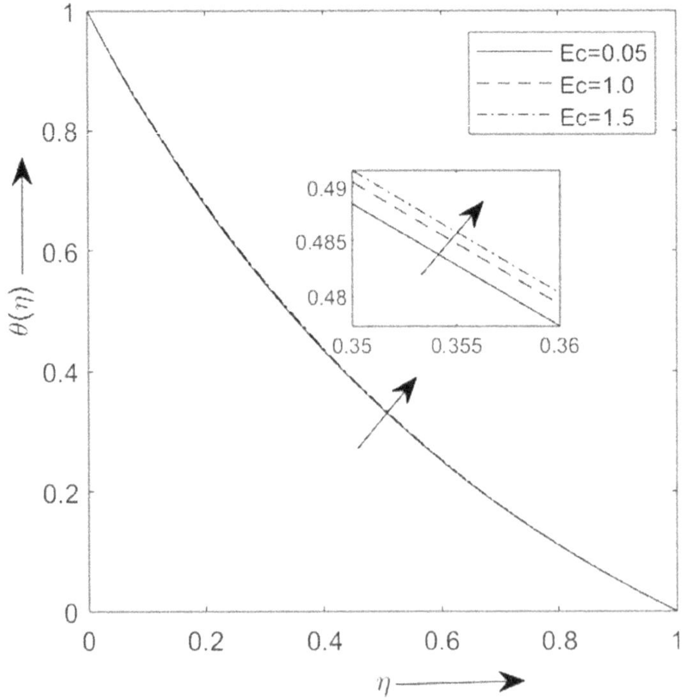

Figure 7.8 T.P with the alteration of *Ec* when $\mathcal{A} = 1$, $Pr = 1.0$, $S_P = 0.4$, $M = 0.5$, $N = 1$.

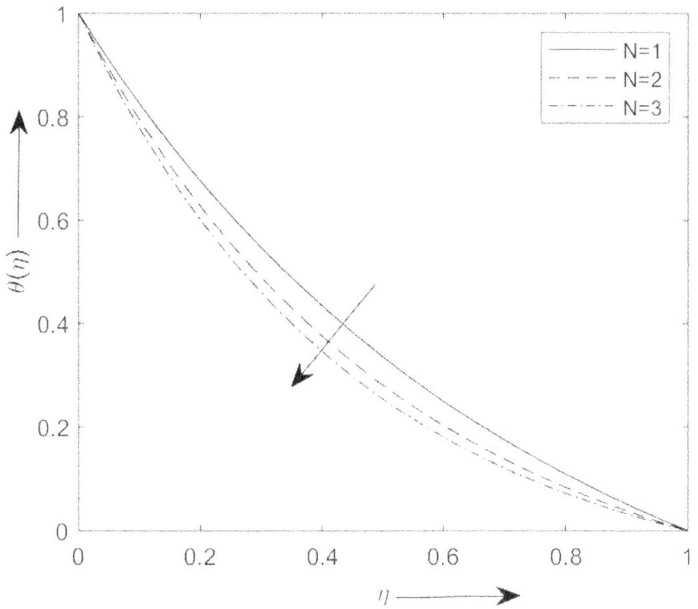

Figure 7.9 T.P with the alteration of N when $\mathcal{A} = 1$, $M = 0.5$, $Pr = 1.0$, $S_P = 0.4$, $Ec = 0.05$.

Table 7.1 Skin friction when $M = 0.5$, $S_P = 0.4$, $Pr = 1.0$, $Ec = 0.05$

$\mathcal{A}$	$f''(0)$
1	−1.8335
2	−2.0621
3	−2.2741

Table 7.2 Nusselt number when $M = 0.5$, $S_P = 0.4$, $Pr = 1.0$, $Ec = 0.05$

$\mathcal{A}$	$\theta'(0)$
1	−1.2691
2	−1.3915
3	−1.5088

Figures 7.5 and 7.6, respectively, show the fluctuation in the temperature profile with various values for M and $\mathcal{A}$. It is spotted that the temperature increases and declines with the variation in the values of M and $\mathcal{A}$, respectively. As $\mathcal{A}$ escalates, Nu's magnitude $|-\theta'(0)|$ rises at the surface which is depicted in Table 7.2.

The temperature profile and the impact of Pr are depicted in Figure 7.7. It's observed when Pr rises and temperature falls, indicating a relationship between the velocity and thermal boundary layer thickness. It's perceived that large values of Pr infer that the thermal diffusivity is predominant and hence thermal boundary layer is narrower than that of the velocity boundary layer. In fact, as surface distance rises, the temperature declines and approaches zero asymptotically in the zone of free stream flow.

The result of Ec on temperature is depicted in Figure 7.8. The fluid's temperature is observed to rise as the Eckert number rises, which physically denotes a fall in the fluid's enthalpy and an increase in its kinetic energy.

Figure 7.9 shows how radiation has an impact on (T.P) temperature profiles. It is noticed that when N rises, the temperature of the fluid falls.

7.5 CONCLUSION

The findings are translated into the following conclusions, which are given below:

- With a rise in Hartmann number (M) and unsteady parameter ($\mathcal{A}$), the velocity profile at the surface decreases, respectively.
- With a rise in the porosity parameter (S_P), the velocity profile declines, but it happens gradually.
- With a rise in Hartmann number (M) and unsteady parameter ($\mathcal{A}$), the temperature profile at the surface increases and declines, respectively.
- With the rise in Prandtl number (Pr), the temperature decreases, indicating that the viscous boundary layer is wider than the thermal boundary layer.
- As Eckert number (Ec) increases, the fluid's temperature increases.
- The temperature diminishes at the surfaces with the rise in radiation parameter (N).
- Due to a rise in unsteady parameter ($\mathcal{A}$), the skin friction and Nusselt number magnitude increases.

NOMENCLATURE

a, b, c	Constant
$\mathcal{A}$	Unsteady parameter (c/a),
B_0	Constant magnetic field, (N m/A)
C_P	Specific heat at constant pressure, (J/kgK)
f	Dimensionless stream function
Ec	Eckert number
k_0	Permeability of porous medium, (m^2)
M	Magnetic parameter (Hartmann number)

N	Radiation parameter
Nu	Nusselt number
Pr	Prandtl number
q_r	radiative heat flux, (W/m^2)
S_P	Porosity parameter
t	Dimensionless time, (K)
T	Fluid's temperature, (K)
u	Fluid's velocity along x-direction, (m/s)
v	Fluid's velocity along y-direction, (m/s)
(x, y)	Cartesian coordinates

GREEK SYMBOLS

ρ	Density of fluid, (kg/m^3)
μ	Dynamic viscosity, (Pa s)
σ_e	Electrical conductivity, $(1/\Omega\text{m})$
η	Dimensionless similarity variable
ν	Kinematic viscosity, (m^2/s)
κ	Thermal conductivity, (W/mK)
ψ	Stream function
τ	Skin friction
θ	Dimensionless temperature

SUPERSCRIPT

' With regard to η, differentiation

SUBSCRIPT

w	Values at the plate
∞	Conditions at the free stream

REFERENCES

Ahmad, N., Siddiqui, Z., and Mishra, M. Boundary layer flow and heat transfer past a stretching plate with variable thermal conductivity. *International Journal of Non-linear Mechanics* 45 (04 2010), 306–309.

Alarifi, I., Abo-Khalil, A., Osman, M., Lund Baloch, L., Mossaad, B. A., Belmabrouk, H., and Tlili, I. MHD flow and heat transfer over vertical stretching sheet with heat sink or source effect. *Symmetry* 11 (02 2019), 2–14.

Andersson, H. Mhd flow of a viscoelastic fluid past a stretching surface. *Acta Mechanica* 95, 1–4 (1992), 227–230.

Andersson, H. I., Aarseth, J. B., and Dandapat, B. S. Heat transfer in a liquid film on an unsteady stretching surface. *International Journal of Heat and Mass Transfer* 43, 1 (2000), 69–74.

Ariel, P. D., Hayat, T., and Asghar, S. Homotopy perturbation method and axisymmetric flow over a stretching sheet. *International Journal of Nonlinear Sciences and Numerical Simulation* 7, 4 (2006), 399–406.

Banks, W. Similarity solutions of the boundary-layer equations for a stretching wall. *Journal de M'ecanique th'eorique et appliqu'ee* 2, 3 (1983), 375–392.

Choudhary, M. K., Chaudhary, S., and Sharma, R. Unsteady MHD flow and heat transfer over a stretching permeable surface with suction or injection. *Procedia Engineering* 127 (2015), 703–710. *International Conference on Computational Heat and Mass Transfer (ICCHMT) -2015.*

Crane, L. J. Flow past a stretching plate. *Zeitschrift f'ur angewandte Mathematik und Physik ZAMP* 21 (1970), 645–647.

Dharmendar Reddy, Y., Shankar Goud, B., Nisar, K. S., Alshahrani, B., Mahmoud, M., and Park, C. Heat absorption/generation effect on MHD heat transfer fluid flow along a stretching cylinder with a porous medium. *Alexandria Engineering Journal* 64 (2023), 659–666.

Elbashbeshy, E. M. Heat transfer over a stretching surface with variable surface heat flux. *Journal of Physics D: Applied Physics* 31, 16 (1998), 1951.

Elbashbeshy, E. M., and Aldawody, D. A. Heat transfer over an unsteady stretching surface with variable heat flux in the presence of a heat source or sink. *Computers & Mathematics with Applications* 60, 10 (2010), 2806–2811.

Ghaly, A. Y. Radiation effects on a certain MHD free-convection flow. *Chaos, Solitons & Fractals* 13, 9 (2002), 1843–1850.

Ishak, A. Unsteady MHD flow and heat transfer over a stretching plate. *Journal of Applied Sciences* 10 (12 2010), 2127–2131.

Ishak, A., et al. MHD boundary layer flow due to an exponentially stretching sheet with radiation effect. *Sains Malaysiana* 40, 4 (2011), 391–395.

Jhankal, A. K., Jat, R. N., and Kumar, D. Unsteady MHD flow and heat transfer over a porous stretching plate. *International Journal of Computational and Applied Mathematics*, 2 (2017), 325–333.

Megahed, A. M., Ghoneim, N. I., Reddy, M. G., and El-Khatib, M. Magnetohydrodynamic fluid flow due to an unsteady stretching sheet with thermal radiation, porous medium, and variable heat flux. *Advances in Astronomy* 2021 (01 2021), 6686–6883.

Raptis, A., Perdikis, C., and Takhar, H. Effect of thermal radiation on MHD flow. *Applied Mathematics and Computation* 153, 3 (2004), 645–649.

Rosca, N. C., and Pop, I. Unsteady boundary layer flow over a permeable curved stretching/shrinking surface. *European Journal of Mechanics - B/Fluids* 51 (2015), 61–67.

Sakiadis, B. Boundary-layer behavior on continuous solid surfaces: II. The boundary layer on a continuous flat surface. *AIChE Journal* 7, 2 (1961a), 221–225.

Sakiadis, B. C. Boundary-layer behavior on continuous solid surfaces: I. Boundary-layer equations for two-dimensional and axisymmetric flow. *AIChE Journal* 7, 1 (1961b), 26–28.

Shah, S. A. G. A., Hassan, A., Karamti, H., Alhushaybari, A., Eldin, S. M., and Galal, A. M. Effect of thermal radiation on convective heat transfer in MHD boundary layer carreau fluid with chemical reaction. *Scientific Reports* 13, 1 (03 2023), 4117.

Siri, Z., Ghani, N. A. C., and Kasmani, R. M. Heat transfer over a steady stretching surface in the presence of suction. *Boundary Value Problems* 2018, 1 (08 2018), 126.

Wang, C. The three-dimensional flow due to a stretching flat surface. *The Physics of Fluids* 27, 8 (1984), 1915–1917.

Chapter 8

Modelling of cylindrical blast waves with dust particles

S. D. Ram
SBP Government Polytechnic Azamgarh

Dhanpal Singh and Ekta Jain
University of Delhi

Mithilesh Singh
Purvanchal University

8.1 INTRODUCTION

In the environment of the Earth, blast waves or detonation waves often propagate. They happen when massive amounts of energy are abruptly concentrated in a small area; common examples include earthquakes, nuclear explosions, supernovas and coal-mine explosions. Due to its wide-ranging applications in the coal-mine explosion, star formation, supersonic flight in polluted air and other science and engineering problems, the formation of strong shock waves in dusty gas also sparked great interest among scientists and researchers in many fields, such as nuclear science, astrophysics and geophysics. It has been found that pressure variations across the shock and other flow characteristics when a blast wave propagates through a gaseous medium that contains a significant amount of dust particles are significantly different from those that occur when the shock passes through the medium without dust particles. In the papers authored by Vishwakarma et al. [1] and Chadha and Jena [2], the volume and mass fraction of solid particles in dusty gases are thought to be relatively small, making it possible to ignore the interactions between the individual particles.

To set up the entire class of self-similar solutions to the problem of shock wave propagation in a dusty gas, a group theoretic approach is used by [3]. An axisymmetric dusty gas flow and a weak discontinuity's interaction with a shock wave have both been studied in [4]. Through the use of an analytical approach [5], investigated weak shock waves in generalised geometry and were able to precisely resolve the problem at hand. As per reference [6], the investigation focuses on the phenomenon of a collapsing cylindrical shock wave under the influence of a magnetic field with infinite electrical conductivity, as explored in previous studies [7–10], and [11] examined the shock wave

 DOI: 10.1201/9781032712079-8

propagation issue while taking the gas molecules' co-volume into account, [12] has found an analytical solution to the problem of an unsteady one-dimensional self-similar flow field between a powerful shock and a moving piston behind it in a dusty gas, [13] used a series expansion method with respect to the inverse square of the shock Mach number to examine the impact of suspended particles on cylindrical blast waves. Recent research on the one-dimensional Riemann issue for the unsteady planar flow of a polytropic/isentropic inviscid compressible fluid in the presence of dust particles has been done by [14–16]. By [17–19], they have already discussed the essential idea of quasi-similarity as well as the validity of the techniques. A number of problems related to strong shock waves with effects of dust particles and also consideration of non-idealness have been investigated by [20–23].

The powerful shock wave problem in the domain of dusty gas has been solved in the current chapter using the notion of quasi-similarity. Furthermore, it is expected that the particles are of very small size and are spaced out very closely throughout the region. The key tenet of quasi-similarity theory is that at the shock front, the distribution of velocity, density and pressure is essentially undetectable. The distribution of the flow characteristics was determined using a numerical calculation based on this supposition. We compare the computational solution produced by the Runge–Kutta method of fourth order with a quasi-similar solution.

8.2 GOVERNING MODELS

The basic formulations for the one-dimensional unsteady flow of the strong shock wave problem can be stated as [24,25]

$$\rho_t + u\rho_x + \rho u_x + \frac{\rho u}{x} = 0, \tag{8.1}$$

$$u_t + u u_x + \frac{1}{\rho} p_x = 0, \tag{8.2}$$

$$p_t + up_x + c^2\rho\left(u_x + \frac{u}{x} \right) = 0, \tag{8.3}$$

where subscripts denote partial derivative, u is the velocity, ρ is the density, p is the pressure of the gas, the entity $c = \sqrt{\Gamma p\left(\rho(1-Z)\right)^{-1}}$ is the speed of sound affected by dust particles, $Z = V_{sp}V^{-1}$ is the volume fraction of the solid particles with V_{sp} and V volume concentration of solid particle and gas, respectively, $\Gamma = \gamma(1+\lambda\omega)(1+\lambda\omega\gamma)^{-1}$ with $\lambda = k_p(1-k_p)^{-1}$, $\omega = c_{sp}\, c_p^{-1}$, c_{sp} is the specific heat of the solid particle, $\gamma = c_p / c_v$ is the specific heat ratio of the gas, c_p is the specific heat of the gas at constant pressure, c_v is the specific heat of the gas at constant volume, $k_p = m_{sp} / m_g$ represents mass fraction of the solid particles, m_{sp} is the mass of the solid particle, and m_g is the mass of the gas. t is the time, and x is the distance from the centre being radial in cylindrical symmetric flows.

The governing equations (8.1)–(8.3) is associated with the equation of state [11,14,26,27]

$$p = \left(\frac{1-k_p}{1-Z}\right)\Re\rho T, \tag{8.4}$$

where $\Re$ and T are the gas constant and the temperature, respectively. The mass fraction of the solid particles at the equilibrium state must be constant; therefore, $\frac{Z}{\rho} = \text{const.}$ (θ say), where $\theta = k_p\, \rho_{sp}^{-1}$ with ρ_{sp} is the density of the solid particles. If we set $k_p = 0$ in equation (8.4) (i.e. the gas is free from the dust particles), then we have $\Gamma = \gamma$, and equation (8.4) takes to the equation of state for an ideal gas.

8.3 THE RANKINE–HUGONIOT RELATIONS

The Rankine–Hugoniot (R–H) relations at the shock are governed by the rule of conservation of mass, momentum and energy across the shock front $x = R(t)$ are given as [9]

$$u\Big|_{x=R} = \frac{2U(1-Z_0)}{(\gamma+1)}\left(1-\frac{1}{M^2}\right), \tag{8.5}$$

$$\frac{p}{p_0}\Bigg|_{x=R} = \frac{2\Gamma}{(\Gamma+1)}M^2 - \frac{(\Gamma-1)}{(\Gamma+1)}, \tag{8.6}$$

$$\frac{\rho}{\rho_0}\Bigg|_{x=R} = \frac{(\Gamma+1)M^2}{2(1-Z_0)+(\Gamma-1+2Z_0)M^2}, \tag{8.7}$$

where $M = U / c$ is the Mach number of the shock front $U = dR/dt$ is the shock speed and R radius of shock front at the core surface $x = R^*$, R^* is the radius of the centre core, $Z_0 = \theta\,\rho_0$ subscripts '0' refer to the value ahead of the shock front. The entities Z_0 and k_p are also related as $Z_0 = k_p\left(k_p + G(1-k_p)\right)^{-1}$ with $G = \rho_{sp}\,\rho^{-1}$.

8.4 TRANSFORMATION OF THE FLOW PARAMETERS

The system of equations (8.1)–(8.3) is transformed into a non-dimensional form by the use of additional new variables $\zeta(x,t) = x/R$ and $y(t) = R/R_0$, where the length of characteristic R_0 in the disturbed region will be $(E/2\pi p_0)^{1/2}$.

The framework of the flow field is represented by the parameters $g(\zeta, y)$, $h(\zeta, y)$ and $f(\zeta, y)$, which we are now going to introduce

$$\left.\begin{aligned} g(\zeta,y) &= p(\zeta,y)/M^2 p_0 \\ h(\zeta,y) &= \rho(\zeta,y)/\rho_0, \\ f(\zeta,y) &= u(\zeta,y)/U. \end{aligned}\right\} \tag{8.8}$$

Using relation (8.8) to convert the governing equations (8.1)–(8.3) into a non-dimensional form, we have

$$y\frac{\partial f}{\partial y} + (f-\zeta)\frac{\partial f}{\partial \zeta} + \frac{1}{\gamma h}\frac{\partial g}{\partial \zeta} + bf = 0, \tag{8.9}$$

$$y\frac{\partial h}{\partial y} + (f-\zeta)\frac{\partial h}{\partial \zeta} + h\left(\frac{\partial f}{\partial \zeta} + \frac{f}{\zeta}\right) = 0, \tag{8.10}$$

$$y\frac{\partial g}{\partial y} + (f-\zeta)\frac{\partial g}{\partial \zeta} + \frac{\gamma g}{(1-Z_0 h)}\left(\frac{\partial f}{\partial \zeta} + \frac{f}{\zeta}\right) + 2bg = 0, \tag{8.11}$$

where $b = \dfrac{y}{U}\dfrac{\partial U}{\partial y}$.

Using the relation $\eta = 1/M^2$, the Rankine–Hugoniot relations (8.5)–(8.7) are transformed as

$$\left.\begin{aligned} g(\zeta,y)\big|_{\zeta=1} &= \frac{2\Gamma-(\Gamma-1)\eta}{(\Gamma+1)}, \\ h(\zeta,y)\big|_{\zeta=1} &= \frac{(\Gamma+1)}{(\Gamma-1+2Z_0)+2(1-Z_0)\eta}, \\ f(\zeta,y)\big|_{\zeta=1} &= \frac{2(1-Z_0)(1-\eta)}{(\Gamma+1)}, \end{aligned}\right\} \tag{8.12}$$

$$f(\zeta_0, y) = \zeta_0. \tag{8.13}$$

In the medium of dusty gas, the internal energy carried by shock wave per unit area is given by ([17–18,25])

$$E = \int_{\zeta_0 R}^{R} \left(\frac{1}{2} \rho u^2 + \frac{(p - p_0)}{(\Gamma - 1)} (1 - Z) \right) dV, \tag{8.14}$$

where $\zeta_0 R = R^*$ and dV is the volume element in the disturbed region. If we take $\zeta_0 = 0$ then internal energy $E = \text{const.}$

Also, the equation (8.14) can be transformed by using relation (8.12) as follows:

$$\frac{1}{y} = \sqrt{M^2 J - \frac{(1 - \zeta_0^2)}{2(\Gamma - 1)}}, \tag{8.15}$$

with

$$J = \int_{\zeta_0}^{1} \left(\frac{1}{2} \Gamma (1 + Z_0) h f^2 + \frac{g}{(\Gamma - 1)} \left(1 + Z_0 h + \frac{Z_0 h}{g M^2} \right) \right) \zeta \, d\zeta. \tag{8.16}$$

8.5 THEORY OF QUASI-SIMILARITY

Concept of quasi-similarity According to [9,17,18], the shock strengths have characteristics that are similar to the flow parameter distribution. Consequently, it is carried out close to the shock front and could be used as

$$f = X(\zeta) Y(y)$$

using the previous relationship as a guide and differentiating the above relationship with regard to y, we have

$$\frac{\partial f}{\partial M} = X(\zeta) \frac{\partial Y(y)}{\partial M}$$
$$= f \times \text{function of } y.$$

The values of the flow parameter f, g and h at $\zeta = 1$ are given by the jump relations (8.5)–(8.7). Thus, $\frac{1}{Y}\left(\frac{dY}{dM}\right)$ may be given in term of M as

$$(\log(f))_R = \frac{2}{(M^2 - 1)} \frac{b}{R}, \tag{8.17}$$

$$(\log(g))_R = \frac{2(\Gamma - 1)}{\Gamma(2M^2 - 1) + 1} \frac{b}{R}, \tag{8.18}$$

$$\left(\log(h)\right)_R = \frac{4(1-Z_0)}{M^2(\Gamma-1+2Z_0)+2(1-Z_0)}\frac{b}{R}. \tag{8.19}$$

Using the above relation (8.17)–(8.19), the system of equations (8.9)–(8.11) is modified into the system of ordinary differential equations as follows:

$$\frac{df}{d\zeta} = \frac{1}{(\zeta-f)}\left[\frac{1}{\Gamma h}\frac{dg}{d\zeta} + \frac{\left(M^2+1\right)}{\left(M^2-1\right)}bf\right], \tag{8.20}$$

$$\frac{dh}{d\zeta} = \frac{h}{(\zeta-f)}\left[\frac{df}{d\zeta} + \frac{4(1-Z_0)}{M^2(\Gamma-1+2Z_0)+2(1-Z_0)}b + \frac{f}{\zeta}\right], \tag{8.21}$$

$$\frac{dg}{d\zeta} = \frac{g}{(\zeta-f)}\left[\frac{\Gamma}{(1-Z_0h)}\frac{df}{d\zeta} + \frac{4\Gamma M^2}{\left(\Gamma\left(2M^2-1\right)+1\right)}b + \frac{1}{(1-Z_0h)}\frac{\Gamma f}{\zeta}\right]. \tag{8.22}$$

The above relations (8.20)–(8.22) are known as quasi-similar solutions.

For equations (8.20)–(8.22), the adiabatic integral is given by

$$h^{(\Omega\Gamma)}\left[(\zeta-f)h\zeta\right]^{\frac{4\Gamma b\left[\frac{M^2}{\left(\Gamma\left(2M^2-1\right)+1\right)}-\frac{4\Omega}{(\Gamma-1)M^2+2}\right]}{2+\left\{4b(1-Z_0)/\left((\Gamma-1+2Z_0)M^2+2(1-Z_0)\right)\right\}}} = \frac{(\Gamma+1)M^2 g}{\left((2M^2-1)\Gamma+1\right)}\left[\frac{1-\Omega}{Z_0}\right]^{(\Omega\Gamma)}, \tag{8.23}$$

where the constant Ω is given as

$$\Omega = \left[1 + \frac{(\Gamma+1)M^2 Z_0}{(\Gamma-1+2Z_0)M^2+2(1-Z_0)}\right].$$

To determine the decay coefficient b with the help of relation (8.15), and with the help of energy relation, we have

$$\frac{1}{p_0 R}\frac{dE}{dR} = 2MJR\frac{dM}{dR} + 2M^2 J + M^2 R\frac{dJ}{dR} - \frac{\left(1-\zeta_0^2\right)}{\Omega(\Gamma-1)}. \tag{8.24}$$

Moreover, we get a relation between the decay coefficient b and variation of energy, given by

$$b = -1 + \frac{1}{2p_0 R M^2 J}\left(\frac{dE}{dR}\right) - \frac{R}{2J}\left(\frac{dJ}{dR}\right) + \frac{1}{2M^2 J\Omega}\frac{\left(1-\zeta_0^2\right)}{(\Gamma-1)}. \tag{8.25}$$

In case of a strong shock wave, the above relation is reduced as

$$b = -1 - \frac{R}{2J}\frac{dJ}{dR} + \frac{1}{2M^2 J\Omega}\frac{\left(1-\zeta_0^2\right)}{(\Gamma-1)}. \tag{8.26}$$

Using the system of equations (8.17)–(8.19) and differentiation equation (8.16) with respect to R, we get

$$\frac{R}{b}\frac{dJ}{dR} = \left\{\frac{4(1-Z_0)}{\left((\Gamma-1+2Z_0)M^2+2(1-Z_0)\right)} + \frac{4}{(M^2-1)}\right\}\int_{\zeta_0}^{1}\frac{\Gamma}{2(1-Z_0)}hf^2\zeta\, d\zeta$$
$$+\frac{1}{\Omega}\left(\frac{2}{(2M^2-1)\Gamma+1}\right)\int_{\zeta_0}^{1} g\zeta\, d\zeta. \tag{8.27}$$

Using the above relation in equation (8.26) and taking the assumption of quasi-similarity yield

$$b = -1 - \frac{1}{2J}\left[\left\{\frac{4(1-Z_0)}{\left((\Gamma-1+2Z_0)M^2+2(1-Z_0)\right)} + \frac{4}{(M^2-1)}\right\}\int_{\zeta_0}^{1}\frac{b\Gamma}{2(1-Z_0)}hf^2\zeta\, d\zeta\right.$$
$$\left.+\left(\frac{2b}{(2M^2-1)\Gamma+1}\right)\frac{1}{\Omega}\int_{\zeta_0}^{1} g\zeta\, d\zeta - \frac{1}{M^2\Omega}\frac{\left(1-\zeta_0^2\right)}{(\Gamma-1)}\right] \approx \delta. \tag{8.28}$$

For any shock strength, we use the following method to obtain the answer while applying the notions of quasi-similarity. To compute the value of b from equation (8.28) by using the results of the foregoing process, we first choose a random sample value of b and use it in the system of equations (8.20)–(8.22) and integrate with conditions (8.12). Normally, δ and b do not agree, so we must take another particular value of b and repeat the process until δ and b are equal. The value of the decay coefficient b, which is provided by Table 8.1, is calculated using the aforementioned approach up to six decimal places.

8.6 ANALYTICAL SOLUTION

The analytical solution is improved by the addition of two more variables, each of which is defined as

$$\left.\begin{aligned} \phi &= (x-f), \\ \psi &= 1-\frac{\phi^2 h}{\Lambda g}, \end{aligned}\right\} \tag{8.29}$$

Table 8.1 Variation of decay coefficient b across the shock front with parameter k_p, G & ω

$\gamma = 1.4,\ \omega = 0.80,\ k_p = 0.40$				$\gamma = 1.4,\ G = 100,\ k_p = 0.40$			$\gamma = 1.4,\ \omega = 0.80,\ G = 100$		
M	$G = 50$	$G = 100$	$G = 1000$	$\omega = 0.0$	$\omega = 0.5$	$\omega = 1.0$	$k_p = 0.0$	$k_p = 0.20$	$k_p = 0.40$
2.0	−0.676876	−0.666140	−0.655786	−0.652339	−0.660911	−0.669492	−0.642857	−0.650937	−0.666140
2.2	−0.707409	−0.696669	−0.686261	−0.686690	−0.692418	−0.699569	−0.677438	−0.683434	−0.696669
2.5	−0.741466	−0.730631	−0.720062	−0.727088	−0.728243	−0.732531	−0.718135	−0.720908	−0.730631
3.0	−0.782249	−0.771335	−0.760578	−0.777447	−0.772049	−0.771498	−0.769039	−0.767295	−0.771335
4.0	−0.838149	−0.827819	−0.817457	−0.844831	−0.832550	−0.825451	−0.837766	−0.831246	−0.827819
5.0	−0.876668	−0.867552	−0.858284	−0.887354	−0.873678	−0.864141	−0.881633	−0.873973	−0.867552
∞	−1.000000	−1.000000	−1.000000	−1.000000	−1.000000	−1.000000	−1.000000	−1.000000	−1.000000

with $\Lambda = (1 - Z_0 h)^{-1}$.

Adding the aforementioned variables φ and ψ to the system of equations (8.20)–(8.22), we obtain

$$\frac{d\phi}{dx} = A_{11} + \left(A_{22} - \delta\frac{\phi}{x}\right)\frac{1}{\psi} + A_{33}\frac{x}{\phi}\left(1 - \frac{1}{\psi}\right), \tag{8.30}$$

$$\frac{d\psi}{dx} = \frac{(1-\psi)}{\phi}\left\{A_{44} - (\Lambda\Gamma + 1)\frac{d\phi}{dx} - \delta(\Lambda\Gamma - 1)\frac{\phi}{x}\right\}, \tag{8.31}$$

The values of the constants A_{11}, A_{22}, A_{33} and A_{44} are given as

$$A_{11} = \left(\frac{1+\eta}{1-\eta}\right)b + 1, \quad A_{22} = \delta + b\left[\frac{4}{\Lambda\left((1-\Gamma)\eta + 2\Gamma\right)} - \frac{1+\eta}{1-\eta}\right], \quad A_{33} = -A_{11} + 1,$$

$$A_{44} = (\Lambda\Gamma - 1)(1+\delta) - b\left[\frac{4(1-Z_0)\eta}{2(1-Z_0)\eta + (\Gamma - 1 + 2Z_0)} - \frac{4\Gamma}{(1-\Gamma)\eta + 2\Gamma}\right].$$

Further, the R–H relation reduces as

$$\phi(1) = \frac{(\Gamma - 1 + 2Z_0) + 2(1-Z_0)\eta}{(\Gamma + 1)}, \tag{8.32a}$$

$$\psi(1) = 1 - \frac{(\Gamma - 1 + 2Z_0) + 2(1-Z_0)\eta}{\left((1-\Gamma)\eta + 2\Gamma\right)\Lambda}, \tag{8.32b}$$

$$\phi(x_0) = 0, \quad \psi(x_0) = 1. \tag{8.32c}$$

As [27] for the closed form solution taking $Z << 1$ so that $(1 - Z_0 h) \approx (1 - Z_0)$.

8.6.1 A general relationship

It's clear from the summary of the calculations' outcomes that the solution may be substantially represented by a relation [17–18]

$$\phi = \beta_{11}(x - x_0) + \beta_{22}(x - x_0)^m, \tag{8.33}$$

where the entity β_{11}, β_{22} and m are obtained by using asymptotic behaviour of ϕ and ψ to $x = x_0$ with initial conditions, we have

$$\beta_{11} = \left(\frac{A_{11} + A_{22}}{2}\right), \quad \beta_{22} = \frac{\phi(1) - \beta_{11}(1 - x_0)}{(1 - x_0)^m}, \quad m = \frac{(\phi'(1) - \beta_{11})(1 - x_0)}{\phi(1) - \beta_{11}(1 - x_0)}.$$

By incorporating equation (8.30) into (8.31) and subsequently integrating the resulting equation, we obtain

$$\psi = 1 + \left(\psi(1) - 1\right)\left(\frac{\phi}{\phi(1)}\right)^{1+\Lambda\Gamma+\frac{A_{44}}{(m-1)\beta_{11}}}\left(\frac{x - x_0}{1 - x_0}\right)^{-\frac{mA_{44}}{(m-1)\beta_{11}}} x^{(\Lambda\Gamma-1)}. \tag{8.34}$$

The integration constant is determined by the relations (8.32). Similarity, from equations (8.20)–(8.22), we have

$$\left(\frac{h}{h(1)}\right) = \left(\frac{\phi(1)}{\phi}\right)^{1+\left(2+\frac{4b(1-Z_0)}{2(1-Z_0)+(\Gamma-1+2Z_0)M^2}\right)\frac{1}{(m-1)\beta_{11}}}\left(\frac{x - x_0}{1 - x_0}\right)^{\left(2+\frac{4b(1-Z_0)}{2(1-Z_0)+(\Gamma-1+2Z_0)M^2}\right)\frac{m}{(m-1)\beta_{11}}} x^{-1}, \tag{8.35}$$

$$\left(\frac{g}{g(1)}\right) = \left(\frac{\phi(1)}{\phi}\right)^{\left(\frac{m+\delta}{m-1}\right)\Lambda\Gamma}\left(\frac{x - x_0}{1 - x_0}\right)^{\frac{\Gamma m(1+\delta)\delta}{(m-1)}} x^{-\Lambda\Gamma}. \tag{8.36}$$

The nature of flow variables $f(\zeta, y)$, $g(\zeta, y)$, and $h(\zeta, y)$ in the disturbed region are shown in Figures 8.1–8.9.

8.7 RESULT AND DISCUSSION

For the proposed problem, the distribution of the flow parameters behind the shock front is obtained by the numerical integration of equations (8.20)–(8.22) with the boundary conditions (8.12), by using Runge–Kutta method of the fourth order. For the above numerical and analytical computation,

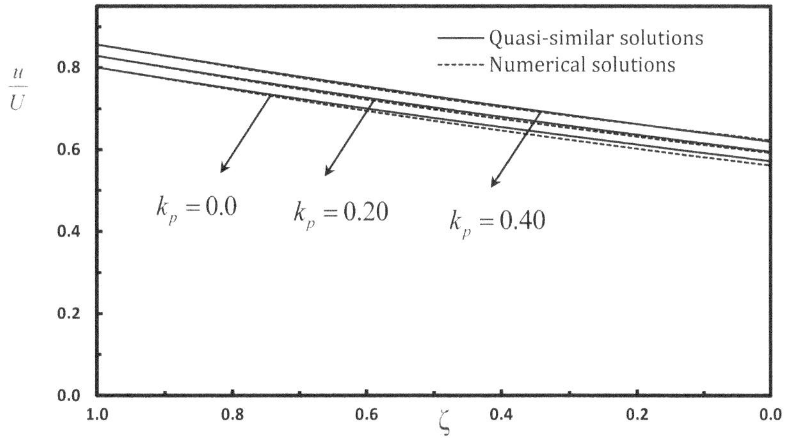

Figure 8.1 Velocity variations $M = 5.0$, $\omega = 0.8$, $G = 100$ for the different values of k_p.

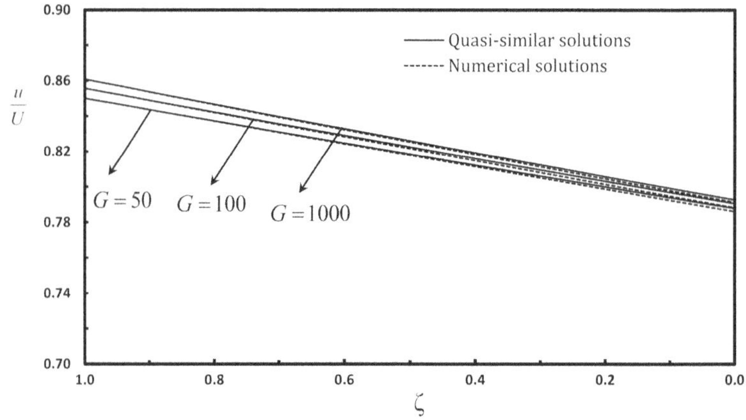

Figure 8.2 Velocity variations $M = 5.0$, $\omega = 0.8$, $k_p = 0.40$ for the different values of G.

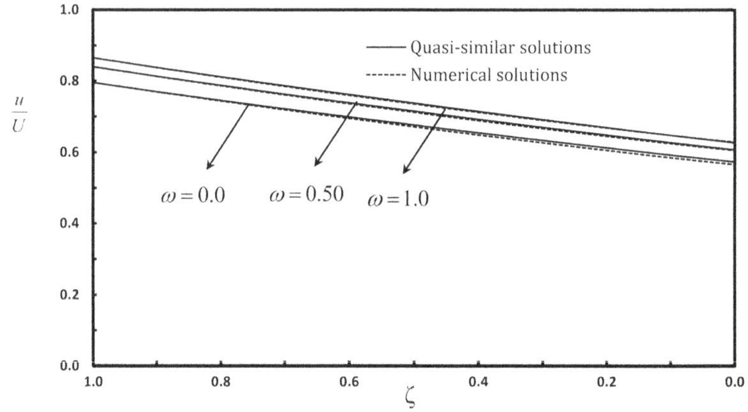

Figure 8.3 Velocity variations $M = 5.0$, $G = 100$, $k_p = 0.40$ for the different values of ω.

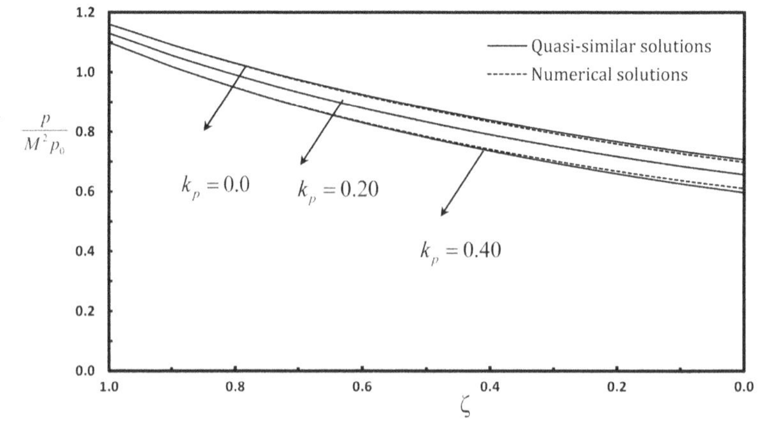

Figure 8.4 Pressure variations $M = 5.0$, $\omega = 0.8$, $G = 100$ for the different values of k_p.

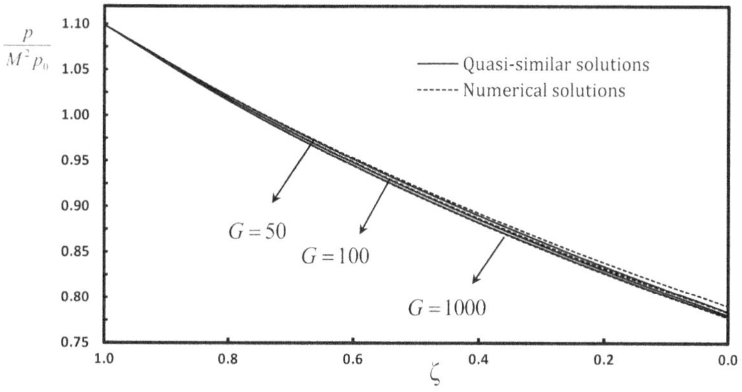

Figure 8.5 Pressure variations $M = 5.0$, $\omega = 0.8$, $k_p = 0.40$ for the different values of G.

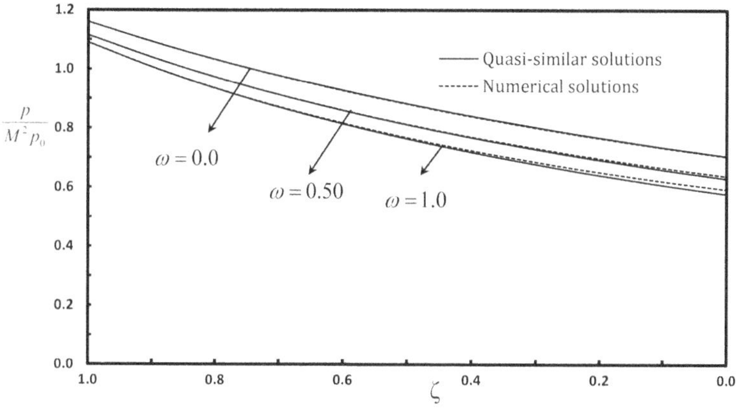

Figure 8.6 Pressure variations $M = 5.0$, $G = 100$, $k_p = 0.40$ for the different values of ω.

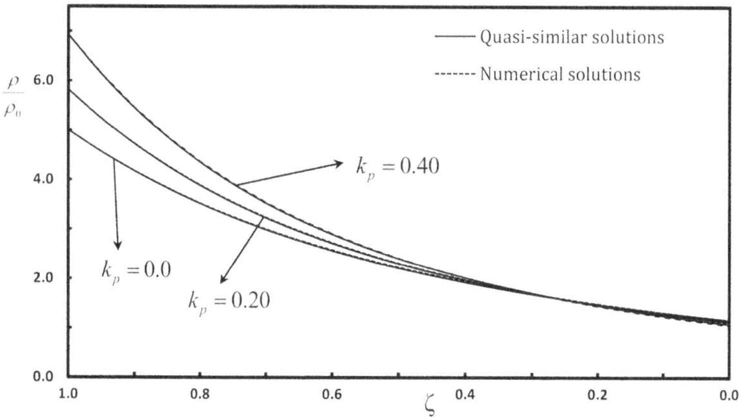

Figure 8.7 Density variations $M = 5.0$, $\omega = 0.8$, $G = 100$ for the different values of k_p.

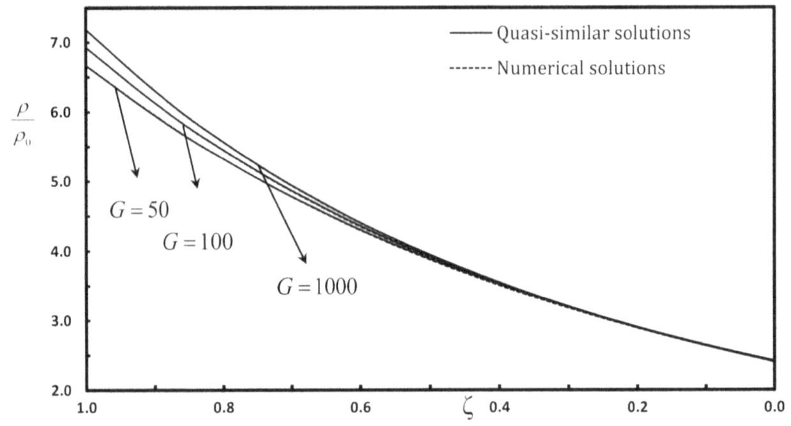

Figure 8.8 Density variations $M = 5.0$, $\omega = 0.8$, $k_p = 0.40$ for the different values of G.

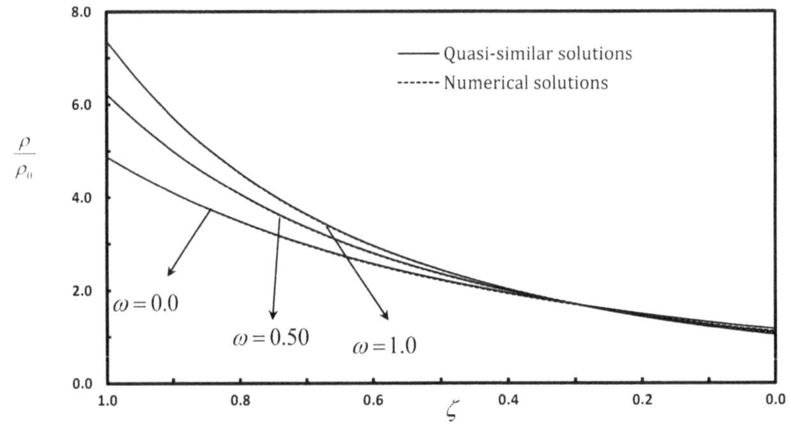

Figure 8.9 Density variations $M = 5.0$, $G = 100$, $k_p = 0.40$ for the different values of ω.

the typical values of the physical quantities are taken as [26], $50 \le G \le 1000$, $\gamma = 1.4$, $0.0 \le k_p \le 0.4$, $2 \le M \le \infty$, $\omega = 0.80$.

Table 8.1 shows the impact of the parameter k_p on the numerically computed decay coefficient "b". For low Mach numbers, it is noticed that when parameter k_p values grow, the decay coefficient "b" marginally increases. Every time a higher Mach number is reached, the process is somewhat sped up. In addition, it is stable at very high Mach numbers. With $G = 50$; 100 and 1000, the competed value of "b" is a drop for low Mach numbers and an increase for higher Mach numbers. The characteristics of the decay coefficient for parameter ω matched the impact of G.

Figures 8.1–8.9 illustrate the behaviour of the flow parameters behind the sock front. Quasi-similar solutions are represented as solid lines, while numerical solutions calculated using the Runge–Kutta fourth-order method are given as dotted lines. The ideal gas situation, where $k_p = 0$, corresponds to fluid that is clearly devoid of dust particles. As the parameters k_p, G and ω are increased, the velocity profile at the shock front $\zeta = 1$ increases. As the shock front is moved further away, the velocity profile gradually decreases, as shown in Figures 8.1–8.3. With higher values of the parameters k_p, G and ω, the density profile behind the shock front exhibits an exponential decline in the Figures 8.4–8.6. However, at the shock front with $\zeta = 1$, significant modifications are visible. With the parameters k_p and ω, it can be shown that the pressure profile grows at the shock front while decreasing with the value G. As demonstrated in Figures 8.7–8.9, the increase of the pressure profile exhibits a reversible relationship with the distance from the shock front.

We deduced from our computation that the compressible fluid has become unstable due to the growing values of the parameters k_p, G and ω. It can also be observed that at the tiny radius of the shock front, the analytical solution, computed using the quasi-similarity method, closely matches both numerical solutions and experimental data. We can draw the conclusion that the quasi-similarity method provides a very accurate approximation in the dusty fluids.

8.8 CONCLUSION

The notion of quasi-similarity has been used to investigate how the influence of dust particles on the flow field behaviour of intense shock waves in an ideal gas. Up to six decimal places have been used in computing the decay coefficient's numerical value. It is clear that the parameters k_p, ω and G influence the decay coefficient. Based on the current analysis, we discover that the flow parameters velocity, pressure and density in the disturbed zone are affected by the parameters k_p, ω and G. The absolute values of velocity and pressure rise at the shock front and progressively fall as we go away from it due to an increase in the parameters k_p and ω, but the density profile rises at the shock front and falls off exponentially as we move away from it. The growth of the velocity and density profiles first occurs at the shock front, while the growth of the pressure profile declines as we travel away from the shock front in accordance with the effects of the parameters k_p and. This method's validity has already been covered by [9,17,18,19].

REFERENCES

[1] Vishwakarma, J. P., Nath, G., &Srivastava, R. K. (2018). Self-similar solution for cylindrical shock waves in a weakly conducting dusty gas. *Ain Shams Engineering Journal*, 9(4), 1717–1730.

[2] Chadha, M., & Jena, J. (2014). Self-similar solutions and converging shocks in a non-ideal gas with dust particles. *International Journal of Non-Linear Mechanics*, *65*, 164–172.
[3] Jena, J., & Sharma, V. D. (1999). Self-similar shocks in a dusty gas. *International Journal of Non-Linear Mechanics*, *34*(2), 313–327.
[4] Conforto, F. (2001). Interaction between weak discontinuities and shocks in a dusty gas. *Journal of Mathematical Analysis and Applications*, *253*(2), 459–472.
[5] Singh, L. P., Ram, S. D., & Singh, D. B. (2011). Exact solution of planar and nonplanar weak shock wave problem in gasdynamics. *Chaos, Solitons & Fractals*, *44*(11), 964–967.
[6] Singh, L. P., Ram, S. D., & Singh, D. B. (2013). The influence of magnetic field upon the collapse of a cylindrical shock wave. *Meccanica*, *48*, 841–850.
[7] Singh, L. P., Ram, S. D., & Singh, D. B. (2011). Analytical solution of the blast wave problem in a non-ideal gas. *Chinese Physics Letters*, *28*(11), 114303.
[8] Singh, M., Husain, A., & Singh, L. P. (2012). Converging shock wave in a dusty gas through nonstandard analysis. *Ain Shams Engineering Journal*, *3*(3), 313–319.
[9] Singh, L. P., Ram, S. D., & Singh, D. B. (2012). Quasi-similar solution of the strong shock wave problem in non-ideal gas dynamics. *Astrophysics and Space Science*, *337*, 597–604.
[10] Singh, L. P., Singh, R., & Ram, S. (2012). Growth and decay of acceleration waves in non-ideal gas flow with radiative heat transfer. *Central European Journal of Engineering*, *2*, 418–424.
[11] Chaudhary, J. P., Ram, S. D., & Singh, L. P. (2019). The plane piston problem with weak gravitational field in a dusty gas. *Journal of King Saud University-Science*, *31*(4), 1027–1033.
[12] Steiner, H., &Hirschler, T. (2002). A self-similar solution of a shock propagation in a dusty gas. *European Journal of Mechanics-B/Fluids*, *21*(3), 371–380.
[13] Higashino, F., & Suzuki, T. (1980). The effect of particles on blast waves in a dusty gas. *Zeitschrift für Naturforschung A*, *35*(12), 1330–1336.
[14] Nath, T., Gupta, R. K., & Singh, L. P. (2017). Solution of Riemann problem for ideal polytropic dusty gas. *Chaos, Solitons & Fractals*, *95*, 102–110.
[15] Gupta, R. K., Nath, T., & Singh, L. P. (2016). Solution of Riemann problem for dusty gas flow. *International Journal of Non-Linear Mechanics*, *82*, 83–92.
[16] Chauhan, A., &Arora, R. (2020). Solution of the Riemann problem for an ideal polytropic dusty gas in magnetogasdynamics. *Zeitschrift für Naturforschung A*, *75*(6), 511–522.
[17] Oshima, K. (1962). Blast waves produced by exploding wires. In *Exploding Wires: Volume 2 Proceedings of the Second Conference on the Exploding Wire Phenomenon*, Boston, November 13–15, 1961, under the Sponsorship of the Geophysics Research Directorate, Air Force Cambridge Research Laboratories, Office of Aerospace Research, with the Cooperation of the Lowell Technological Institute Research Foundation. pp. 159–174, Springer US.
[18] Oshima, K. (1964). *Quasisimilar Solutions of Blast Waves.* Aeronautical Research Institute.

[19] Abdel-Raouf, A. M., &Gretler, W. (1991). Quasi-similar solutions for blast waves with internal heat transfer effects. *Fluid Dynamics Research*, *8*(5–6), 273.
[20] Arora, R., Tomar, A., & Pal Singh, V. (2012). Similarity solutions for strong shocks in a non-ideal gas. *Mathematical Modelling and Analysis*, *17*(3), 351–365.
[21] Anand, R. K. (2014). On dynamics of imploding shock waves in a mixture of gas and dust particles. *International Journal of Non-Linear Mechanics*, *65*, 88–97.
[22] Chauhan, A., Arora, R., &Siddiqui, M. J. (2019). Propagation of blast waves in a non-ideal magnetogasdynamics. *Symmetry*, *11*(4), 458.
[23] Sharma, K., Arora, R., Chauhan, A., &Tiwari, A. (2020). Propagation of waves in a nonideal magnetogasdynamics with dust particles. *Zeitschrift für Naturforschung A*, *75*(3), 193–200.
[24] Whitham, G. B. (1974). *Linear and Nonlinear Waves*. John Wiley & Sons.
[25] Courant, R. (1948). K. O. Gas Dynamics. Friedrichs. *Supersonic Flow and Shock Waves*, 1, 143–156. John Wliey & sons.
[26] Pai, S. I. (1977). *Two-Phase Flows* (Vol. 3). Springer-Verlag.
[27] Pai, S. I., Menon, S., & Fan, Z. Q. (1980). Similarity solutions of a strong shock wave propagation in a mixture of a gas and dusty particles. *International Journal of Engineering Science*, *18*(12), 1365–1373.

Chapter 9

Reliable analysis of Riemann problem in magnetogasdynamics

Mithilesh Singh
VBS, Purvanchal University

Shakuntla Sharma
Tula's Institute

Nidhi Handa
Kanya Gurukula Mahavidyalaya

9.1 INTRODUCTION

Magnetohydrodynamics (MHD) is applied to nuclear science, fusion research and many astrophysical and geophysical problems and to problems of fusion power where the application is the creation and containment of hot plasmas by electromagnetic forces, since material walls would be destroyed. Several of the fluid properties are studied in the presence of a magnetic field. We can see the study of Euler equation of hydrodynamics to MHD equations by Li (2005), he supposed a new MHD-HLLC solver same like Toro and Batten's HLLC solver in pure HD limit and satisfying the conservation laws in MHD.

Zachary et al. (1994) studied the higher-order Godunov method for the solution of two and three-dimensional equations of ideal MHD; his work was based on both suitable operator split approximation to the full multidimensional equations and one-dimensional Riemann solver. Shen (2011) examined Riemann solution of the isentropic MGD equations that converges to the corresponding Riemann solution of the transport equations by letting both pressure and magnetic field vanish. Delmont et al. (2009) discussed the classical problem of planar shock rare fraction at an oblique density discontinuity, separating two gases at rest. Powell (1997) analyzed an approximate Riemann solver for the governing equations of ideal MHD, his Riemann solver was based on eight-wave structure. Many of the authors have studied Riemann problem of conventional gas dynamics (Courant and Friedriches 1990; Glimm 1965; Lax 1957; Smoller 1994; Toro 1997).

 DOI: 10.1201/9781032712079-9

A study of Riemann's problem for quasilinear hyperbolic system of equations, the governing equations of the one-dimensional unsteady flow of an inviscid and perfectly conducting gas subjected to transverse magnetic field solved by Singh and Singh (2014). Liu and Sun (2013) also analyzed the Riemann problem for one-dimensional ideal isentropic magnetogasdynamics with transverse magnetic field and for the existence and uniqueness of the solution of the Riemann problem by the characteristic method. Cuong and Thanh (2015) used a Godunov-type scheme for the solution of non-conservative, isentropic model of a fluid flow in a nozzle with variable cross-section.

Hu and Sheng (2013) discussed the Riemann problem for one-dimensional magnetogasdynamics in Lagrangian coordinates using the characteristic analysis method. Dai and Woodward (1995) investigated a simple approximate Riemann solver for a hyperbolic system of conservation laws to use in the Godunov scheme. The Riemann problem and elementary wave interaction for the unsteady simple flow of an inviscid, isentropic and perfectly conducting fluid conditional on a transverse magnetic field is discussed by Sekhar and Sharma (2010). Li et al. (2020) studied the Riemann Problem and Vanishing Magnetic Field Limit. Wancheng et al. (2022) analyzed the Riemann problem for isentropic magnetogasdynamics in a variable cross-section duct.

In this paper, we studied elementary waves and their related discontinuities in the region influenced by traverse magnetic field; here we have analyzed the behavior of fluid in terms of density, velocity and pressure of the fluid crosswise the shock waves, rarefaction wave and contact discontinuity in the presence of the magnetic field.

9.2 GOVERNING EQUATIONS

Governing equations for a conservative form of one-dimensional for quasilinear hyperbolic system of equations conditional on the transverse magnetic field in magnetogasdynamics, Singh et al. (2010) are:

$$\rho_t + u\rho_r + \rho u_r = 0, \tag{9.1}$$

$$u_t + uu_r + \rho^{-1}(p_r + \mu BB_r) = 0, \tag{9.2}$$

$$p_t + up_r + a^2 \rho u_r = 0, \tag{9.3}$$

$$B_t + uB_r + Bu_r = 0, \tag{9.4}$$

where $\rho(r,t)$, $u(r,t)$, $p(r,t)$ and $B(r,t)$ denote the density, particle velocity, pressure and magnetic field, respectively, where $a = (\gamma p / \rho)^{1/2}$ is the speed of sound with γ being the constant specific heat ratio, μ being the magnetic permeability, where r is the single spatial co-ordinate and t is the time. The subscripts denote the partial differential equation unless stated otherwise.

This system of equations (9.1)–(9.4) can be written as in conservative form:

$$\frac{\partial}{\partial t}\begin{bmatrix} \rho \\ u \\ p \\ B \end{bmatrix} + \begin{bmatrix} u & \rho & 0 & 0 \\ 0 & u & 1/\rho & \mu B/\rho \\ 0 & a^2\rho & u & 0 \\ 0 & B & 0 & u \end{bmatrix} \frac{\partial}{\partial r}\begin{bmatrix} \rho \\ u \\ p \\ B \end{bmatrix} = \begin{bmatrix} 0 \\ 0 \\ 0 \\ 0 \end{bmatrix}. \tag{9.5}$$

Equation (9.5) can be written as

$$U_t + MU_r = 0, \tag{9.6}$$

where U is the vector of conserved variables as follows:

$$U = \begin{pmatrix} \rho \\ u \\ p \\ B \end{pmatrix}, \quad M = \begin{bmatrix} u & \rho & 0 & 0 \\ 0 & u & 1/\rho & \mu B/\rho \\ 0 & a^2\rho & u & 0 \\ 0 & B & 0 & u \end{bmatrix}, \tag{9.7}$$

where M is the Jacobian flux matrix which determines the main features of the flow and wave interaction curves whose eigenvalues are as follows:

$$\lambda^{(1)} = u - \eta, \quad \lambda^{(2,4)} = u, \quad \lambda^{(3)} = u + \eta, \tag{9.8}$$

where $\eta = (a^2 + b^2)^{1/2}$ is the effective speed of sound and $b^2 = \mu B^2 / \rho$ is the square of Alfven speed. Eigen vectors of the matrix M are

$$r^{(1)} = \begin{bmatrix} \rho/\eta \\ 1 \\ \rho\eta \\ 0 \end{bmatrix}, \quad r^{(2)} = \begin{bmatrix} 1 \\ 0 \\ 0 \\ 0 \end{bmatrix}, \quad r^{(3)} = \begin{bmatrix} -\rho/\eta \\ 1 \\ -\rho/\eta \\ 0 \end{bmatrix}, \quad r^{(4)} = \begin{bmatrix} 1 \\ 0 \\ -b^2 \\ 1 \end{bmatrix}. \tag{9.9}$$

The system (9.6) is written in conservative flux form

$$\frac{\partial U}{\partial t}+\frac{\partial F(U)}{\partial r}=0, \tag{9.10}$$

where

$$U=\begin{bmatrix}\rho\\ \rho u\\ B\end{bmatrix},\quad F(\mathrm{U})=\begin{bmatrix}\rho u\\ \rho u^2+p+\mu B^2/2\\ uB\end{bmatrix}, \tag{9.11}$$

with the assumption $B=k\rho$ (Shen 2011), where k is positive constant using this condition system (9.10) is equivalent to

$$U_t^*+M^*U_r^*=0, \tag{9.12}$$

where

$$U^*=\begin{pmatrix}\rho\\ u\\ p\end{pmatrix},\quad M^*=\begin{bmatrix}u & \rho & 0\\ b^2/\rho & u & 1/\rho\\ 0 & a^2\rho & u\end{bmatrix}. \tag{9.13}$$

Eigenvalues and eigenvectors of matrix M^* is given by

$$\lambda^{(1)}=u-\eta,\quad \lambda^{(2)}=u,\quad \lambda^{(3)}=u+\eta, \tag{9.14}$$

$$r^{(1)}=\begin{bmatrix}-\rho/\eta\\ 1\\ -\rho a^2/\eta\end{bmatrix},\quad r^{(2)}=\begin{bmatrix}1\\ 0\\ -b^2\end{bmatrix},\quad r^{(3)}=\begin{bmatrix}\rho/\eta\\ 1\\ a^2\rho/\eta\end{bmatrix}. \tag{9.15}$$

Eigenvalues are divided into two groups, one having a characteristic field as linearly degenerate and the other having a characteristic field as genuinely nonlinear. The second characteristic field is linearly degenerate since $\nabla\lambda^{(2)}r^{(2)}=0$, there is an indication of contact discontinuity; characteristic field corresponding to first and third eigenvalues is genuinely nonlinear; the discontinuities arise here is either in the form of shock waves or rarefaction waves.

The system of equations (9.10) has been written as

$$\rho_t+(\rho u)_r=0, \tag{9.16}$$

$$(\rho u)_t + (\rho u^2 + p + \mu B^2 / 2)_r = 0, \tag{9.17}$$

$$B_t + (uB)_r = 0. \tag{9.18}$$

Rankine–Hugoniot relation is as follows:

$$\sigma\left(U - U_l\right) = F(U) - F(U_l). \tag{9.19}$$

Using equation (9.19) in equations (9.16)–(9.18), we have

$$\sigma[\rho] = [\rho u], \tag{9.20}$$

$$\sigma[\rho u] = [\rho u^2] + [p] + \mu[B^2] / 2, \tag{9.21}$$

$$\sigma[B] = [uB], \tag{9.22}$$

where $\left[\rho\right] = \rho_r - \rho_l$, etc. is the jump across the discontinuity and the shock speed σ of the discontinuity is satisfying equations (9.20)–(9.22), we get

$$\sigma = \left[\frac{1}{\left(\rho - \rho_0\right)}\left\{\rho u^2 - \rho_0 u_0^2 + k\left(\rho^\gamma - \rho_0^\gamma\right) + \frac{\mu}{2}\left(\rho^2 - \rho_0^2\right)\right\}\right]^{1/2}. \tag{9.23}$$

The Hugoniot set $H(U_0)$ consisting of all right-hand states $U(\rho, u, p)$ connected to all left-hand states $U_0(\rho_0, u_0, p_0)$ by a shock, which gives Hugoniot curves as given by

$$H_1(U_0): u = u_0 \pm (\rho - \rho_0)\left[\frac{1}{\rho\rho_0}\left\{\frac{\left(\rho^\gamma - \rho_0^\gamma\right)}{\left(\rho - \rho_0\right)} + \frac{\mu k^2}{2}\left(\rho + \rho_0\right)\right\}\right]^{1/2}. \tag{9.24}$$

For eigenvectors $r_i(U_0)$, $i = 1, 3$; i-Hugoniot curves are

$$H_1(U_0): u = u_0 - (\rho - \rho_0)\left[\frac{1}{\rho\rho_0}\left\{\frac{\left(\rho^\gamma - \rho_0^\gamma\right)}{\left(\rho - \rho_0\right)} + \frac{\mu k^2}{2}\left(\rho + \rho_0\right)\right\}\right]^{1/2}, \tag{9.25}$$

$$H_3(U_0): u = u_0 + (\rho - \rho_0)\left[\frac{1}{\rho\rho_0}\left\{\frac{\left(\rho^\gamma - \rho_0^\gamma\right)}{\left(\rho - \rho_0\right)} + \frac{\mu k^2}{2}\left(\rho + \rho_0\right)\right\}\right]^{1/2}. \tag{9.26}$$

By any one of the flow variables including the shock speed we can determine the Hugoniot curves; if $\rho > \rho_0$, then for the eigenvector r_1, the Hugoniot curve $H_1(U_0)$ strictly decreases and the Hugoniot curve $H_3(U_0)$ associated with the third eigenvector strictly increases.

9.3 SHOCK WAVES

A shock wave enters into a region of lower pressure across, which pressure and density jump to higher values and all the primitive variables involved are discontinuous. Let us denote by $S_i(U_0)$ the forward curves of admissible i-shock waves which consist of all right-hand states U that can be connected to a given left state U_0 by an i-lax shock wave $i = 1,3$ corresponding with first and third eigenvectors associated with genuinely nonlinear characteristic field.

$$S_1(U_0): u = u_0 - (\rho - \rho_0)\left[\frac{(1)}{\rho\rho_0}\left\{\frac{(\rho^\gamma - \rho_0^\gamma)}{(\rho - \rho_0)} + \frac{\mu k^2}{2}(\rho + \rho_0)\right\}\right]^{1/2}, \quad \rho \geq \rho_0, \tag{9.27}$$

$$S_3(U_0): u = u_0 - (\rho - \rho_0)\left[\frac{(1)}{\rho\rho_0}\left\{\frac{(\rho^\gamma - \rho_0^\gamma)}{(\rho - \rho_0)} + \frac{\mu k^2}{2}(\rho + \rho_0)\right\}\right]^{1/2}, \quad \rho \leq \rho_0. \tag{9.28}$$

Similarly, backward shock waves $S_i^B(U_0)$ consisting of all left-hand states U that can be connected to a given right-hand state U_0 by an i-Lax shock wave, $i = 1,3$ are given by

$$S_1^B(U_0): u = u_0 + (\rho - \rho_0)\left[\frac{(1)}{\rho\rho_0}\left\{\frac{(\rho^\gamma - \rho_0^\gamma)}{(\rho - \rho_0)} + \frac{\mu k^2}{2}(\rho + \rho_0)\right\}\right]^{1/2}, \quad \rho \leq \rho_0, \tag{9.29}$$

$$S_3^B(U_0): u = u_0 + (\rho - \rho_0)\left[\frac{(1)}{\rho\rho_0}\left\{\frac{(\rho^\gamma - \rho_0^\gamma)}{(\rho - \rho_0)} + \frac{\mu k^2}{2}(\rho + \rho_0)\right\}\right]^{1/2}, \quad \rho \geq \rho_0. \tag{9.30}$$

For the genuinely nonlinear characteristic field, the forward shock wave curves $S_1(U_0)$ and $S_3(U_0)$ strictly decrease and backward shock wave curves $S_1^B(U_0)$ and $S_3^B(U_0)$ strictly increase due to the effect of the magnetic field.

9.4 CURVE OF RAREFACTION WAVES

A rarefaction wave is a discontinuity, which is moving in the opposite direction of gas flow and all the variables involved are continuous. This wave propagates with constant speed in the (x,t) plane, and the solution obtained $U(x,t)$ is a constant solution that is independent of rescaling of x-axis.

The solution of equation (9.12) in the variable ξ is given by

$$U(x,t) = V(\xi), \quad \xi = x/t, \quad x \in R, \quad t > 0. \tag{9.31}$$

Substituting $U(x,t) = V(\xi)$ in equation (9.12), we get

$$\left(A(V(\xi)) - \xi I\right)V'(\xi) = 0, \quad \xi = x/t. \tag{9.32}$$

Using equation (9.32), we get the differential equation as

$$V'(\xi) = \frac{r_i(V(\xi))}{\nabla\lambda_i V(\xi) \cdot r_i(V(\xi))}, \quad i = 1,3. \tag{9.33}$$

The function $U(x,t)$ is the solution of system of equations (9.32) in variable ξ

$$U(x,t) = \begin{cases} U_l & x < \lambda_i(U_l)t \\ V\left(\dfrac{x}{t}\right) & \lambda_i(U_l)t \le x \le \lambda_i(U_r)t. \\ U_r & x > \lambda_i(U_r)t \end{cases} \tag{9.34}$$

The denominator of (9.33) does not vanish for the eigenvectors r_1, r_3, and then we have the integral curves

$$\rho(\xi) = \left(\rho_l^{\frac{\gamma-1}{2}} + \frac{(-1)^i \varepsilon^2(\gamma-1)}{(\gamma+\varepsilon^2)\sqrt{\gamma k}}(\xi - \lambda_i(U_i))\right)^{\frac{2}{\gamma-1}}, \tag{9.35}$$

$$u(\xi) = u_l + \frac{2\varepsilon^2}{\gamma+\varepsilon^2}(\varepsilon - \lambda_i(U_i)). \tag{9.36}$$

Let U and U_0 denote the equilibrium states of the fluid separating the region of the rarefaction wave curve, the forward-rarefaction wave curves of all right states U can be connected to a given left state U_0 by an i-rarefaction wave, $i = 1,3$ corresponding to the first and third characteristic values associated with the genuinely nonlinear characteristic field, which gives 1-CRW (centered rarefaction waves) and 3-C RW are:

$$R_1(U_0): u = u_0 - \frac{2\sqrt{\gamma k}}{\gamma - 1}\left(\rho^{\frac{\gamma-1}{2}} - \rho_l^{\frac{\gamma-1}{2}}\right), \quad \rho \leq \rho_0, \tag{9.37}$$

$$R_3(U_0): u = u_0 + \frac{2\sqrt{\gamma k}}{\gamma - 1}\left(\rho^{\frac{\gamma-1}{2}} - \rho_l^{\frac{\gamma-1}{2}}\right), \quad \rho \geq \rho_0, \tag{9.38}$$

the backward i-rarefaction wave for all left-hand state to right-hand state connecting by i-rarefaction wave $i = 1,3$ as follows:

$$R_1^B(U_0): u = u_0 - \frac{2\sqrt{\gamma k}}{\gamma - 1}\left(\rho^{\frac{\gamma-1}{2}} - \rho_l^{\frac{\gamma-1}{2}}\right), \quad \rho \geq \rho_0, \tag{9.39}$$

$$R_3^B(U_0): u = u_0 + \frac{2\sqrt{\gamma k}}{\gamma - 1}\left(\rho^{\frac{\gamma-1}{2}} - \rho_l^{\frac{\gamma-1}{2}}\right), \quad \rho \leq \rho_0. \tag{9.40}$$

There is no effect on rarefaction wave curves when a rarefaction wave passes through a traverse magnetic field. As $\nabla\lambda_i V(\xi)\cdot r_i(V(\xi)) = 1$, here $\gamma = \varepsilon^2$.

In addition, the wave curve which is union of the shock wave curve and rarefaction curve as

$$W_1(U_0) = S_1(U_0) \cup R_1(U_0), \tag{9.41}$$

$$W_2(U_0) = S_2(U_0) \cup R_2(U_0), \tag{9.42}$$

$$W_1^B(U_0) = S_1^B(U_0) \cup R_1^B(U_0), \tag{9.43}$$

$$W_2^B(U_0) = S_2^B(U_0) \cup R_2^B(U_0), \tag{9.44}$$

where

$$W_1(U_0): u = W_1(U_0,\rho) = \begin{cases} u_0 - \dfrac{2\sqrt{\gamma k}}{\gamma-1}\left(\rho^{\frac{\gamma-1}{2}} - \rho_0^{\frac{\gamma-1}{2}}\right), & \rho \le \rho_0 \\ u_0 - (\rho-\rho_0)\left[\dfrac{1}{\rho\rho_0}\left\{ k\dfrac{\left(\rho^\gamma - \rho_0^\gamma\right)}{(\rho-\rho_0)} + \dfrac{\mu k^2}{2}(\rho+\rho_0)\right\}\right]^{1/2}, & \rho \ge \rho_0, \end{cases} \tag{9.45}$$

$$W_3(U_0): u = W_3(U_0,\rho) = \begin{cases} u_0 - \dfrac{2\sqrt{\gamma k}}{\gamma-1}\left(\rho^{\frac{\gamma-1}{2}} - \rho_0^{\frac{\gamma-1}{2}}\right), & \rho \le \rho_0 \\ u_0 + (\rho-\rho_0)\left[\dfrac{1}{\rho\rho_0}\left\{ k\dfrac{\left(\rho^\gamma - \rho_0^\gamma\right)}{(\rho-\rho_0)} + \dfrac{\mu k^2}{2}(\rho+\rho_0)\right\}\right]^{1/2}, & \rho \ge \rho_0, \end{cases} \tag{9.46}$$

$$W_1^B(U_0): u = W_1^B(U_0,\rho) = \begin{cases} u_0 + \dfrac{2\sqrt{\gamma k}}{\gamma-1}\left(\rho^{\frac{\gamma-1}{2}} - \rho_0^{\frac{\gamma-1}{2}}\right), & \rho \le \rho_0 \\ u_0 + (\rho+\rho_0)\left[\dfrac{1}{\rho\rho_0}\left\{ k\dfrac{\left(\rho^\gamma - \rho_0^\gamma\right)}{(\rho-\rho_0)} + \dfrac{\mu k^2}{2}(\rho+\rho_0)\right\}\right]^{1/2}, & \rho \ge \rho_0, \end{cases} \tag{9.47}$$

$$W_3^B(U_0): u = W_3^B(U_0,\rho) = \begin{cases} u_0 + \dfrac{2\sqrt{\gamma k}}{\gamma-1}\left(\rho^{\frac{\gamma-1}{2}} - \rho_0^{\frac{\gamma-1}{2}}\right), & \rho \le \rho_0 \\ u_0 + (\rho-\rho_0)\left[\dfrac{1}{\rho\rho_0}\left\{ k\dfrac{\left(\rho^\gamma - \rho_0^\gamma\right)}{(\rho-\rho_0)} + \dfrac{\mu k^2}{2}(\rho+\rho_0)\right\}\right]^{1/2}, & \rho \ge \rho_0. \end{cases} \tag{9.48}$$

For the genuinely nonlinear characteristic field, curve $W_1(U_0)$ strictly decreases and the curve $W_3^B(U_0)$ strictly increases for both the conditions $\rho \le \rho_0$ and $\rho \ge \rho_0$.

The curve $W_3(U_0)$ strictly decreases for $\rho \le \rho_0$ and strictly increases for $\rho \ge \rho_0$, while the curve $W_1^B(U_0)$ strictly increases for $\rho \le \rho_0$ and strictly decreases for $\rho \ge \rho_0$ after adding a factor of magnetic field.

9.5 STATIONARY WAVES

Here σ is zero and Rankine–Hugoniot jump relation written by equations (9.20)–(9.22) across the shock reduces to the form

$$[\rho u] = 0, \tag{9.49}$$

$$\left[\rho u^2 + p + \frac{\mu}{2} B^2\right] = 0. \tag{9.50}$$

Here $U = (\rho, u, p, B)$ and $U_0 = (\rho_0, u_0, p_0, B_0)$ be the two equilibrium states on both sides of stationary waves satisfying equation (9.10).

It is assumed that U_0 is fixed. The curve $w_3(U_0)$ of stationary waves represented by density function $F(\rho)$, consisting of all the states U that can be connected to U_0 by stationary wave, can be parameterized ρ, using (9.49) in (9.50) we have

$$F(\rho) = \pm\left(\rho\sqrt{\left\{u_0^2 - \mu\left(\rho^{\gamma-1} - \rho_0^{\gamma-1}\right) - 2\mu k^2(\rho - \rho_0)\right\}}\right) = 0, \tag{9.51}$$

$$\frac{\partial F(\rho)}{\partial \rho} = \frac{3\mu k^2 \rho^2 + \frac{\mu(\gamma+1)}{2} - \rho\left(u_0^2 + \mu\rho_0^{\gamma-1} + 2\mu k^2 \rho_0\right)}{\sqrt{u_0^2 - \mu\left(\rho^{\gamma-1} - \rho_0^{\gamma-1}\right) - 2\mu k^2(\rho - \rho_0)}}. \tag{9.52}$$

Further $\dfrac{\partial F(\rho)}{\partial \rho} = 0$ gives

$$A_1\rho^2 + A_2\rho^\gamma + A_3\rho = 0, \tag{9.53}$$

where $A_1 = 3\mu k^2$, $A_2 = \dfrac{\mu(\gamma+1)}{2}$, $A_3 = -\left(u_0^2 + \mu\rho_0^{\gamma-1} + 2\mu k^2\rho_0\right)$.

The equation (9.53) shows that it is nonlinear equation of density in the power of γ. We could not solve easily without using constants. For solving equation (9.53), taking the values of constants as, $\mu = 0.1$, $k = 0.02$ and initial values of flow variables as $u_0 = 0.2$ $\rho_0 = 1.5$ by the Newton-Raphson method and calculated the maximum value of ρ for $1 < \gamma < 2$ as below:

$$\frac{\partial^2 F(\rho)}{\partial \rho^2} = \frac{G_1\rho^{2\gamma-3} + G_2\rho^{\gamma-2} + G_3\rho^{\gamma-1} + G_4\rho + G_5}{\left(u_0^2 - \mu\left(\rho^{\gamma-1} - \rho_0^{\gamma-1}\right) - 2\mu k^2(\rho - \rho_0)\right)^{3/2}} \tag{9.54}$$

where $G_1 = k\gamma(\mu + k\gamma)$, $G_2 = -k\gamma^2\left(u_0^2 + \mu\rho_0^{\gamma-1} + 2\mu k^2\rho_0\right)$, $G_3 = 2k^2 \left(2k^2\gamma^2 + \mu^2\right)$, $G_4 = 3\mu^2 k^4$, $G_5 = -2\mu k^2\left(u_0^2 + \mu\rho_0^{\gamma-1} + 2\mu k^2\rho_0\right)$.

Case1: for $\gamma = 1.4$

$$\frac{\partial F(\rho)}{\partial \rho} > 0, \quad \rho(= 1.68019) < \rho_{max} = 1.6802, \tag{9.55}$$

$$\left(\frac{\partial^2 F(\rho)}{\partial \rho^2}\right)_{\rho=1.68019} = -90.8647. \tag{9.56}$$

The equation (9.56) shows that curve $w_3(U_0)$ is concave and monotonic decreasing.

$$\frac{\partial F(\rho)}{\partial \rho} < 0, \quad \rho(= 3.94799) > \rho_{max} = 1.6802, \tag{9.57}$$

$$\left(\frac{\partial^2 F(\rho)}{\partial \rho^2}\right)_{\rho=3.94799} = 0.0000142174. \tag{9.58}$$

Equation (9.58) shows that the curve $w_3(U_0)$ is convex and monotonic increasing and when magnetic term B is not present there in equation (9.52), $\rho_{max} = 1.68236$; thus, maximum value of ρ decreases when a magnetic term is present there.

Case 2: for $\gamma = 1.67$

$$\frac{\partial F(\rho)}{\partial \rho} > 0, \quad \rho\left(= 1.4082 \times 10^{-9}\right) < \rho_{max} = 1.45335, \tag{9.59}$$

$$\left(\frac{\partial^2 F(\rho)}{\partial \rho^2}\right)_{\rho=1.4082\times10^{-9}} = -7.96615. \tag{9.60}$$

Again, equation (9.60) shows that the curve $w_3(U_0)$ is concave and monotonic decreasing.

$$\frac{\partial F(\rho)}{\partial \rho} < 0, \quad \rho(= 3.24526) > \rho_{max} = 1.45335 \tag{9.61}$$

$$\left(\frac{\partial^2 F(\rho)}{\partial \rho^2}\right)_{\rho=3.24526} = 0.000176095. \tag{9.62}$$

The equation (9.62) shows that curve $w_3(U_0)$ is convex and monotonic increasing and when effect of magnetic field is zero in (9.52) then value of density is maximum at $\rho_{max} = 1.45404$. Equations (9.59) and (9.62) show that the values of maximum density decrease due to effect of magnetic field.

9.6 DISCUSSION AND CONCLUSION

We have discussed the complete solution of the wave structure of the Riemann problem in an isentropic fluid flow in the presence of a magnetic field at any point (r,t) in the domain assumed with $r_l < r < r_r$ (the subscripts l and r stand for left and right states). For the study of the solution of the system of equations, we found physical features such as wave pattern, characteristics and discontinuities of the fluid in magnetogasdynamics. We have analyzed the rarefaction waves, stationary waves in the presence of an area duct, and shock waves using Rankine–Hugoniot conditions. Further, we have determined the maximum and minimum values of the density function using the Newton-Raphson method, which show that convex and concave curves.

REFERENCES

Courant, R., Friedriches, K.O. (1990). *Supersonic Flow and Shock Waves*. Interscience, New York.

Cuong, D.H., Thanh, M.D. (2015). A Godunov-type scheme for the isentropic model of a fluid flow in a nozzle with variable cross section. *Applied Mathematics and Computation* 256: 602–629.

Dai, W., Woodward, P.R. (1995). A simple Riemann solver and high-order Godunov schemes for hyperbolic system of conservation laws. *Journal of Computational Physics* 121: 51–65.

Delmont, P., Keppens, R., van der Holst, B. (2009). An exact Riemann solver based on solution for regular shock refraction. *Journal of Fluid Mechanics* 627: 33–53.

Glimm, J. (1965). Solution in the large for non-linear hyperbolic system of equations. *Communication on Pure and Applied Mathematics* 18: 697–715.

Hu, Y., Sheng, W. (2013). The Riemann problem of conservation law in magnetogasdynamics. *Communications on Pure and Applied Analysis* 12: 755–769.

Lax, P.D. (1957). Hyperbolic system of conservation law II. *Communication on Pure and Applied Mathematics* 10: 537–566.

Li, S. (2005). An HLLC Riemann Solver for Magnetohydrodynamics. *Journal of Computational Physics* 203: 344–357.

Li, S., Cheng, H., Yang, H. (2020). Pressureless magnetohydrodynamics system: Riemann problem and vanishing magnetic field limit. *Advances in Mathematical Physics* 2020: 13. https://doi.org/10.1155/2020/638254.

Liu, Y., Sun, W. (2013). Riemann problem and wave interactions in magnetogas-dynamics. *Journal of Mathematical Analysis and Applications* 397: 454–466.

Powell, K.G. (1997). *An Approximate Riemann Solver for Magnetogasdynamics.* Springer Verlag, Berlin, Heidelberg.

Raja Sekhar, T., Sharma, V.D. (2010). Riemann problem and elementry wave interactions in sentropic magnetogasdynamics. *Nonlinear Analysis: Real World Application* 11: 619–636.

Shen, C. (2011). The limits of Riemann solution to the isentropic magnetogasdynamics. *Applied Mathematics Letters* 24: 1124–1129.

Singh, L.P., Singh, M., Pandey, B.D. (2010). Analytic solution of converging shock waves in magnetogasdynamics. *AIAA Journal* 48: 2523–2528.

Singh, R., Singh, L.P. (2014). Solution of the Riemann problem in magnetogasdynamics. *International Journal of Non-Linear Mechanics* 67: 326–330.

Smoller, J. (1994). *Shock Waves and Reaction-Diffusion Equations.* Springer-Verlag, New York.

Toro, E.F. (1997). *Riemann Solver and Numerical Methods for Fluid Dynamics.* 2nd edn. Springer-Verliag, Berlin.

Wancheng, S., Xiao, T., Zhang, Q. (2022). Riemann problem for isentropic magneto-gasdynamics in a variable cross-section duct. *Studies in Applied Mathematics* 149: 266–292.

Zachary, A.L., Malgoli, A., Collela, P. (1994). A higher-order method for multidimensional ideal magnetohydrodynamics. *SIAM Journal of Scientific Computing* 15: 263–284.

Chapter 10

Modelling of Rayleigh–Taylor instability with nanoparticles and magnetic field

P. Girotra
RGGC

J. Ahuja
PGGC

10.1 INTRODUCTION

A comprehensive detailed discussion about the instabilities under various assumptions in the field of hydrodynamics and hydromagnetic has been given by Chandrasekhar (1981). However, credit for analysing the placement of heavier fluid on top of lighter fluid goes to Rayleigh (Lord Rayleigh, 1900). The investigation about the type of instability that develops when lighter fluid accelerates against heavier fluid was reported in 1950 (Taylor, 1950). This prevalent event in astrophysics and hydrodynamics is a type of instability known as Rayleigh–Taylor instability (RTI). Since then, numerous theoretical and experimental contributions by illustrious scholars have been documented in this field. Many researchers from the 20th and 21st centuries have been interested in the impact of variables such as surface tension, magnetic field and rotation on the RTI of fluids. The interactions between surface tension and rotation were examined, and it was concluded that the presence of rotation allows the system to stabilise for Rayleigh-Taylor (RT) modes (Sharma et al., 2010). The vertical rotation of the system was looked into, and it was discovered the highest limit of the increasing rate of instability (Chakraborthy, 1979). The effects of rotation using the variational principle were also documented (Hide, 1956). Further, the variational approach was also employed to examine how the magnetic field affected the RT modes of viscous compressible plasma and show how both viscosity and the magnetic field had stabilising effects (Bhatia, 1974). Using Floquet analysis, the combined effects of magnetic field and rotation were observed and the conclusion was made that rotation and a magnetic field cause RTI to be suppressed (Rannacher and Engel, 2007). In addition, the observation about how rotation affected the RTI of magnetised

DOI: 10.1201/9781032712079-10

conducting plasma came to light, and it was discovered that the magnetic field had no bearing on the state of RTI (Sharma and Chhajlani, 1998). When magnetic field was investigated with RTI (Pacitto et al., 2000), they found that the beginning of RTI is delayed as the magnetic field increases. It was theorised that anytime the vertical magnetic field is mixed with heat and mass movement, the magnitude of perturbations (produced by RTI) is minimised (Shukla and Awasthi, 2021). Moreover, the research on the non-linear cases of RTI with magnetic fields was conducted, and it was convincingly documented that a greater magnetic field reduces the non-linear increase of RTI (Carlyle and Hillier, 2017).

In addition, the impact of a magnetic field on RTI has been investigated for several fluid types. While focusing on Oldroydian viscoelastic fluids (Sharma and Bhardwaj, 1994) and Rivlin–Ericksen elastico-viscous fluids (Sharma et al., 2001), the RT instability predominates when magnetic field is not present in the system. It was noted that an oblique magnetic field had a higher impact on fluid RTI than a vertical or horizontal magnetic field in this context (Ariel and Aggarwala, 1979).

The ground-breaking research in the areas of hydrodynamics and hydromagnetic, as it relates to nanofluids, has come to light. Choi was the first to do such experiments (Choi, 1995), and Tzou (2008) studied the role of nanofluids in convective transport, which was then taken up by other researchers. Further researchers investigated the collection of conservation equations that were applied to various models (Nield and Kuznetsov 2014; Bhadauria, 2007). Ahuja and Gupta conducted a thorough examination of the local thermal non-equilibrium model (Ahuja and Gupta, 2019). The impact of the magnetic field on the thermal instability of a nanofluid layer was recorded as the magnetic field stabilises the nanofluid layer in both stationary convection and oscillatory motion (Gupta et al., 2013).

Despite thermal and thermosolutal stability, it is now crucial to study RTI in nanofluids because of the vast array of astrophysical and geophysical applications it possesses. A recent investigation on the RTI in nanofluids was conducted (Ahuja and Girotra, 2021a, 2021b). It was concluded that surface tension helps to stabilise the system but that the Atwood number accelerates the destabilisation process for RT forms of setup. In addition, the configuration to produce the destabilisation is set by the volume fraction of nanoparticles. Following that, some illuminating conclusions regarding the effectiveness of nanoparticles in RTI of rotating nanofluids came into light (Girotra et al., 2021). They proposed that raising the volume fraction of nanoparticles in rotating nanofluids could postpone the onset of RTI in the system.

Recent research demonstrates that RTI is still a topic of interest for scientists and researchers. By resolving the compressible Navier–Stokes equation, a three-dimensional model of RTI at the interface of two air masses is created, in which the influence of thermal gradient on the RTI is investigated by taking into account two temperature differences between the air masses (Sengupta and Verma, 2023). Similar to this, recent research on the effects of thermal gradient stabilisation or destabilisation (Sengupta and Joshi, 2023) with the combined Kelvin–Helmholtz and RTI shows that the initial process is dominated by the vortex stretching or compressibility effect, followed by a sharp increase in baroclinic torque contribution once the buoyancy effects become significant. Shukla and his coworkers recently discovered the influence of heat and mass transfer on the RTI of some non-Newtonian types of fluids, explaining that the instability may be delayed if there is more transmission of heat at the interface (Shukla at el., 2023). It is clear that research is continuing to uncover the hidden facts about RTI, as chemical engineering, biochemical engineering, industry and numerous physical phenomena pertaining to geophysics and astrophysics are all affected by the effect of magnetic field on the Rayleigh–Taylor instability in nanofluids. The problem has applications in geophysics because of the earth's magnetic field and the enhanced properties of nanofluids.

The main objective of this study is to shed light on how the magnetic field significantly affects the RTI reported in two stacked nanofluid layers. Researchers hope to examine this phenomenon's possible applicability in a variety of disciplines by better comprehending it. The researchers in the current chapter use the sophisticated normal mode technique to accomplish this goal. Using this method, they are able to develop a dispersion relation that accounts for a number of crucial elements, including the effects of the magnetic field, surface tension and the volume percentage of nanoparticles inside the nanofluid layers. The RTI's behaviour is described by the dispersion relation, a mathematical equation that takes into account all of the aforementioned affecting factors. The researchers acquire important insights into the intricate interaction of these physical forces and their combined effect on the stability of the nanofluid layers by carefully analysing this relationship. This thorough examination is essential because it reveals the complex mechanisms driving the system's behaviour when the magnetic field and other pertinent factors are present. For researching and maximising possible applications of this phenomenon in diverse technical and scientific fields, it is essential to understand these fundamental processes. By elucidating the magnetic field's effects on the RTI in nanofluid layers, researchers can potentially unlock new possibilities for enhancing fluid mixing, heat transfer and material processing in nanotechnology, microfluids and other related fields. Furthermore, this knowledge may open the door to creative methods for developing and managing nanofluid-based systems for real-world uses including advanced cooling technologies, medication delivery systems and even next-generation energy systems.

10.2 MATHEMATICAL MODEL

In this section, the system under examination consists of two infinite layers of various nanofluids with zero resistivity that are stacked on top of one another and separated by a plane $z = 0$. There are two layers in the region: (a) $z < 0$ as lower layer having density (ρ_1), viscosity (μ_1) and volume fraction of nanoparticles (ϕ_1) and (b) $z > 0$ known as an upper layer having density (ρ_2), viscosity (μ_2) and volume fraction of nanoparticles (ϕ_2). In these two layers of base fluids under discussion, it is presumed that nanoparticles are uniform in size, shape and distribution. The system is experiencing the effect of acceleration due to gravity $\mathbf{g}(0, 0, -g)$ and vertical magnetic field $\boldsymbol{H}(0,0,H)$. In this configuration, surface tension across the interface is effective and it is assumed that nanoparticles are of higher density as compared with base fluids. It is assumed that there is no variation in temperature across the interface. Basic equations (Chandersekhar, 1981; Tzou, 2008) are as follows:

$$\nabla \cdot \boldsymbol{v} = 0, \tag{10.1}$$

$$\boldsymbol{v} \cdot \nabla\phi + \frac{\partial \phi}{\partial t} = \nabla \cdot \left(D_B \nabla \phi\right), \tag{10.2}$$

$$\frac{\partial \rho}{\partial t} = -wD\rho, \tag{10.3}$$

$$\nabla \cdot h = 0, \tag{10.4}$$

$$\frac{\partial h}{\partial t} = \nabla \times (\mathbf{v} \times H), \tag{10.5}$$

$$\rho\left[\boldsymbol{v} \cdot \nabla \boldsymbol{v} + \frac{\partial \boldsymbol{v}}{\partial t}\right] = -\nabla p + \mu \nabla^2 \boldsymbol{v} + \rho \mathbf{g} + \frac{\mu_e}{4\pi}((\nabla \times \boldsymbol{h}) \times \boldsymbol{H}) + \sum\left[T\left(\frac{\partial^2}{\partial x^2} + \frac{\partial^2}{\partial y^2}\right) z_s\right] \delta(z - z_s), \tag{10.6}$$

where $\boldsymbol{v} = \boldsymbol{v}(u, v, w), \boldsymbol{h} = \left(h_x, h_y, h_z\right), \boldsymbol{H} = (0,0,H)$ has components in all three directions.

$$\rho_1 = \phi_1 \rho_{p1} + (1 - \phi_1)\rho_{f1}, \tag{10.7}$$

$$\rho_2 = \phi_2 \rho_{p2} + (1 - \phi_2)\rho_{f2}, \tag{10.8}$$

with the expression

$$\rho = \rho_1 + \rho_2 \tag{10.9}$$

and

$$\rho = \phi \rho_p + (1-\phi)\rho_f. \tag{10.10}$$

10.3 PERTURBATION EQUATIONS

Let $v', \delta\rho, \delta\, p\, \boldsymbol{h}'(h_x, h_y, h_z)$ and $\delta\phi$ denote the perturbations in velocity, density, pressure, magnetic field and volume fraction of nanoparticles respectively. As the system is set for infinitesimal perturbations, therefore the conservation equations (10.1)–(10.6) are transformed into (dashes are dropped for the convenience)

$$\nabla \cdot \boldsymbol{v} = 0, \tag{10.11}$$

$$\frac{\partial(\delta\phi)}{\partial t} = D_B \nabla^2 (\delta\phi), \tag{10.12}$$

$$\frac{\partial(\delta\rho_f)}{\partial t} = -wD\rho_f, \tag{10.13}$$

$$\frac{\partial(\delta\rho_p)}{\partial t} = -wD\rho_p, \tag{10.14}$$

$$\nabla \cdot \boldsymbol{h} = 0, \tag{10.15}$$

$$\frac{\partial \boldsymbol{h}}{\partial t} = H \frac{\partial \boldsymbol{v}}{\partial z}, \tag{10.16}$$

$$\rho \frac{\partial u}{\partial t} - \frac{\mu_e H}{4\pi}\left(\frac{\partial h_x}{\partial z} - \frac{\partial h_z}{\partial x}\right) = -\frac{\partial}{\partial x}(\delta p) + \mu \nabla^2 u, \tag{10.17}$$

$$\rho \frac{\partial v}{\partial t} - \frac{\mu_e H}{4\pi}\left(\frac{\partial h_y}{\partial z} - \frac{\partial h_z}{\partial y}\right) = -\frac{\partial}{\partial y}(\delta p) + \mu \nabla^2 v, \tag{10.18}$$

$$\rho\left[\frac{\partial w}{\partial t}\right] = -\frac{\partial}{\partial z}(\delta p) + \mu \nabla^2 w - g(\delta\rho) + \sum\left[T\left(\frac{\partial^2}{\partial x^2} + \frac{\partial^2}{\partial y^2}\right) z_s\right]\delta(z - z_s). \tag{10.19}$$

Equation (10.10) takes the form as

$$\delta\rho = (\rho_p - \rho_f)\delta\phi + (1-\phi)\delta\rho_f + \phi(\delta\rho_p). \tag{10.20}$$

Using equation (10.20) in equation (10.19) results as

$$\rho\left[\frac{\partial w}{\partial t}\right]=-\frac{\partial}{\partial z}(\delta p)+\mu\nabla^2 w-g\left((\rho_p-\rho_f)\delta\phi+(1-\phi)\delta\rho_f+\phi\delta\rho_p\right)$$
$$+\sum\left[T\left(\frac{\partial^2}{\partial x^2}+\frac{\partial^2}{\partial y^2}\right)z_s\right]\delta(z-z_s). \tag{10.21}$$

In next section, normal modes are used for analysing these equations.

10.4 NORMAL MODE TECHNIQUE

All perturbation equations (10.11)–(10.18) and equation (10.21) are analysed through normal modes with the independent variables x, y and t; expressed as

$$f(z)e^{(ik_x x+ik_y y+nt)}, \tag{10.22}$$

where the parameters such as wave number and growth rate are denoted by $k_1=\sqrt{k_x^2+k_y^2}$ and 'n'. After putting the expression (10.22) in the equations (10.17), (10.18) and (10.21) which gives result as follow:

$$n\rho u-\frac{\mu_e H}{4\pi}(Dh_x-ik_x h_z)=-ik_x(\delta p)+\mu(D^2-k_1{}^2)u, \tag{10.23}$$

$$n\rho v-\frac{\mu_e H}{4\pi}(Dh_y-ik_y h_z)=-ik_y(\delta p)+\mu(D^2-k_1{}^2)v, \tag{10.24}$$

$$n\rho w=-D(\delta p)+\mu\nabla^2\left(D^2-k_1^2\right)w-g\left[(\rho_p-\rho_f)\delta\phi+(1-\phi)\delta\rho_f\right.$$
$$\left.+\phi(\delta\rho_p)\right]+\sum\left[T\left(-k_1^2\right)z_s\right]\delta(z-z_s). \tag{10.25}$$

Further, equations (10.11)–(10.16) will turn as

$$ik_x u+ik_y v=-Dw, \tag{10.26}$$

$$n\delta\phi=\left(D_B(D^2-k_1{}^2)\right)\delta\phi, \tag{10.27}$$

$$n\delta\rho_f=-wD\rho_f, \tag{10.28}$$

$$n\delta\rho_p=-wD\rho_p, \tag{10.29}$$

$$h_z=\frac{H}{n}Dw, \tag{10.30}$$

$$ik_x h_x+ik_y h_y=-Dh_z, \tag{10.31}$$

where 'D' is the differential operator and normal component of velocity 'w_s' is expressed as

$$w_s = nz_s. \tag{10.32}$$

After applying the normal modes, pressure term is supposed to be eliminated from the equations (10.23)–(10.25) with the help of equations (10.27)–(10.32) which leads to the following:

$$\begin{aligned} n\rho(D^2 - k_1^2)w - \frac{\mu_e H^2}{4\pi n}(D^2 - k_1^2)D^2 w = \left[D\left(\mu(D^2 - k_1^2)Dw\right)\right] - k_1^2 \mu(D^2 - k_1^2)w \\ -gk_1^2\left[(\rho_p - \rho_f)cw - \frac{(1-\phi)wD\rho_f}{n} - \frac{w\phi D\rho_p}{n}\right] \\ +\sum Tk_1^4 z_s \delta(z - z_s). \end{aligned} \tag{10.33}$$

Equation (10.33) represents the general relation of RTI of two superimposed nanofluids and in the absence of magnetic field, it coincides with the dispersion relation of Ahuja and Girotra (2021a).

10.5 ANALYTIC SOLUTION AND BOUNDARY CONDITIONS

As we have considered two superimposed nanofluids which are separated by the horizontal boundary at $z = 0$. Let us take viscosity, density and potency of nanoparticles as constant in equation (10.33) which gives

$$\mu\left(D^2 - k_1^2\right)\left(D^2 - q^2\right)w = 0, \tag{10.34}$$

where

$$q^2 = \frac{(k_1^2 + \rho n)}{\mu(1 - H^2/4\pi n^2)}, \tag{10.35}$$

which is defined only if

$$H^2 < 4\pi n^2. \tag{10.36}$$

Since nanofluids have an infinite range of motion, 'w' must disappear in both layers. As a result, equation (10.34) has general solution as follows:

$$w_1 = c_1 e^{k_1 z} + c_2 e^{q_1 z}, \quad \text{for} \quad z < 0, \tag{10.37}$$

$$w_2 = c_3 e^{-k_1 z} + c_4 e^{q_2 z}, \quad \text{for} \quad z > 0, \tag{10.38}$$

where c_1, c_2, c_3, c_4 are constants and q_1, q_2 are defined as $q_i = \sqrt{\frac{(k_1^2 - \rho n)}{4\pi n^2 \mu_i (4\pi n^2 - H^2)}}$ for i=1,2. 'w' in equations (10.37) and (10.38) satisfies some boundary conditions at $z = 0$ like continuity of w, Dw and $\mu\left(D^2 + k_1^2\right)w$, which further lead to

$$c_1 + c_2 = c_3 + c_4, \tag{10.39}$$

$$-k_1c_1 + q_1c_2 = -k_1c_3 - q_2c_4, \tag{10.40}$$

$$\mu_1(2k_1^2c_1 + q_1^2c_2 + k_1^2c_2) = \mu_2(2k_1^2c_3 + q_2^2c_4 + k_1^2c_4). \tag{10.41}$$

Integrating equation (10.33) over the interface while taking the continuity of w, Dw and $\left(D^2 + k_1^2\right)w$ and eliminating the constants from equations (10.39)–(10.41), the following equation is obtained where $q_1 = q_2 = q$

$$\begin{aligned}
&-(\rho_2 Dw_2 - \rho_1 Dw_1) - \frac{\mu_e H^2}{4\pi n^2}(D^2 - k_1^2)(Dw_2 - Dw_1) \\
&= \left[\frac{1}{n}\left\{\mu_2(D^2 - k_1^2)Dw_2 - \mu_1(D^2 - k_1^2)Dw_1\right\}\right] + \frac{gk_1^2}{n^2}, \\
&(\rho_p - \rho_f)cw_0 - \frac{1-\phi}{n^2}gk_1^2(\rho_{f2} - \rho_{f1})w_0 - \frac{\phi}{n^2}gk_1^2 \\
&(\rho_{p2} - \rho_{p1})w_0 + T\frac{w_0k_1^4}{n^2}.
\end{aligned} \tag{10.42}$$

On further simplifying, a dispersion relation is obtained as

$$n^2k_1^2(\rho_1 + \rho_2)\Lambda_1 + n\left[(\mu_2 - \mu_1)\Lambda_1 + \Lambda_2\right] + \frac{H^2\mu_e}{4\pi}\Lambda_3 + \Lambda_4 = 0, \tag{10.43}$$

where

$$\Lambda_1 = 2k_1^2(k_1 + q)(\mu_2 + \mu_1) + k_1^2q^2(\mu_2 + \mu_1) + 2k_1^2q(k_1\mu_1 + q\mu_2), \tag{10.44}$$

$$\begin{aligned}
\Lambda_2 = 2k_1&\left[q(k_1 + q)(q^2 - k_1^2)\mu_2\mu_1 + \mu_1^2q^2(k_1 + q)(k_1 + 2q)\right. \\
&\left.+k_1(q - k_1)\mu_2^2(k_1^2q + q^2 + k_1^2 + k_1q)\right],
\end{aligned} \tag{10.45}$$

$$\Lambda_3 = 4k_1\mu_1(q + k_1)^2 + 2k_1\mu_2q(q - k_1) + 4k_1(q^2 + k_1^2), \tag{10.46}$$

$$\Lambda_4 = -2k_1^2\left\{\mu_2(q^2+k_1^2)+(q+k_1)^2\mu_1\right\}\left[gk_1^4(\rho_p-\rho_f)\phi_0 - \left(g(\rho_2-\rho_1)-Tk_1^{\ 2}\right)\right]. \tag{10.47}$$

In the absence of a magnetic field, the dispersion relation sorted in equation (10.43) coincides with Ahuja and Girotra (2021a). Further, if the limiting case of viscosity is taken, then equation (10.43) confirms the result of Taylor (1950):

$$n^2 = \frac{1}{\rho}\left(-Tk_1^3 - g(\rho_1-\rho_2)\right). \tag{10.48}$$

It is clear as per equation (10.35) that with the condition: $\rho_2 > \rho_1$, all $\Lambda_i; 1 \le i \le 3$ are of positive sign but Λ_4 is of negative sign, therefore Descartes' rule of sign is supported by at least one variation of sign in equation (10.43) which further identifies the onset of the RTI. On substituting the value of 'q' in equation (10.43), the expression comes out to be as

$$\Gamma_0 n^9 + \Gamma_1 n^8 + \Gamma_2 n^7 + \Gamma_3 n^6 + \Gamma_4 n^5 + \Gamma_5 n^4 + \Gamma_6 n^3 + \Gamma_7 n^2 + \Gamma_8 n + \Gamma_9 = 0, \tag{10.49}$$

where $\Gamma_i, 0 \le i \le 9$ are the coefficients. The existence of at least one real root is guaranteed by equation (10.49) which establishes the initial phase of instability. The detailed examination of the functions of several factors, including surface tension, magnetic parameter and nanoparticles' volume fraction, is concluded in the following section.

10.6 DISCUSSION

Here, the updated dispersion relation (equation10.43) is discussed in detail together with modifications to surface tension, magnetic field and nanoparticles' volume fraction. Consider three cases:

1. variations/modifications considered in one parameter at once while maintaining other values constant,
2. variations/modifications considered in two parameters at once while maintaining other values constant,
3. variations/modifications in all three parameters at once.

Such cases are considered to gain important insights on RTI. With appropriate tables and graphs for the system's unstable mode, all of these instances are described and examined. In order to analyse the instability at the interface, the water (ρ_{f1}) and glycerol (ρ_{f2}) with $\left(\rho_{f2} > \rho_{f1}\right)$ are taken as base fluids in lower and upper layer respectively. In base fluid glycerol, aluminium nanoparticles (ρ_{p2}) are dispersed whereas in base fluid water magnesium nanoparticles (ρ_{p1}) are suspended. Numerical values of these parameters are, $\rho_{f1} = 1000\,\text{kg/m}^3$, $\rho_{f2} = 1260\,\text{kg/m}^3$, $\rho_{p1} = 1738\,\text{kg/m}^3$, $\rho_{p2} = 2700\,\text{kg/m}^3$ are taken from Volk and Kahler (2018) and viscosity is $\mu_1 = 8.9\times10^{-4}$ Pa·s and $\mu_2 = 1.41$ Pa·s (Segur and Oberstar, 1951).

Stable Configuration ($\rho_1 > \rho_2$): It is clearly depictable that the system is stable for $\rho_1 > \rho_2$; thus there is no difference in the sign of equation (10.43) for the numerical values discussed and all $\Lambda_1, \Lambda_2, \Lambda_3$ and Λ_4 are of positive sign. Thus, it is stated that the system does not show any perturbation for $\rho_1 > \rho_2$.

Unstable Configuration ($\rho_2 > \rho_1$): For this case, equation (10.43) shows one variation as $\Lambda_4 < 0$ whereas $\Lambda_i > 0; 1 \le i \le 3$ that causes for onset of the instability at $z = 0$. Thus, for **unstable mode** of RTI, the wave number band is sorted for all wave number ranging $0 < k_1 < k_{n1}$ as

$$k_{n1} = \left[\frac{g(\rho_2 - \rho_1)}{T}\right]^{1/2}. \tag{10.50}$$

10.6.1 Variations/modifications considered in one parameter at once

The impact of various parameters has been examined in this section by changing only one parameter while assuming that all others remain constant. 'T' is the only parameter that isn't fixed in Figure 10.1 to examine the impacts of surface tension, and a graph is plotted between wave numbers given by $k\left(=\dfrac{k_1}{10}\right)$ and growth rate is denoted by 'n'. The surface

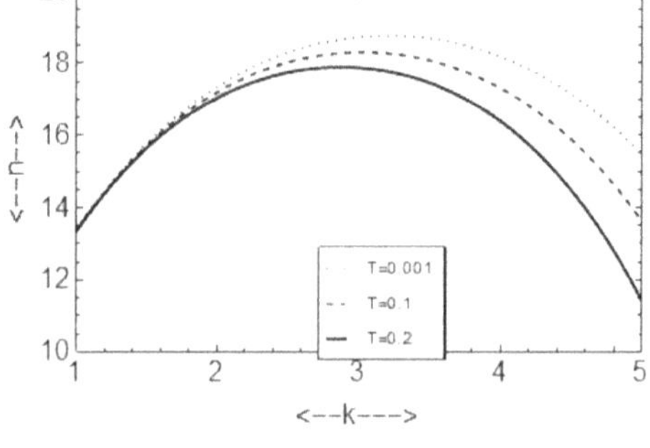

Figure 10.1 Effects of varying values of surface tension on 'n' growth rate under RTI.

tension bears the values as $T = 0.001, 0.1, 0.2$ and the other parameters are supposed to be constant as $\phi = 0.001, H = 1.5$. It is observed that surface tension has a stabilising effect on the configuration for RTI since the critical values of growth rate ('n_c') drop when values of 'T' are raised.

As a consequence, surface tension has a strong stabilising effect (see Table 10.1). Further, Figure 10.2 shows the effect of the magnetic field with $H = 1, 1.5, 2$ while taking other parameters to be constant as $\phi = 0.001, T = 0.1$.

Table 10.1 displays the insightful facts about the critical values 'n_c' of the growth rate when only one parameter is taken as variable quantity while assuring the other parameters to behave as constant.

Once more, the curves with rising values of 'H' exhibit stabilising properties. Therefore, in the unstable mode of setup, the magnetic field aids in delaying the start of RTI. Figure 10.3 illustrates the effect of 'ϕ' with variations as $\phi = 0.001, 0.01, 0.1$ and fixing the values of surface tension and magnetic field as $T = 0.1$, $H = 1.5$. As the critical values of growth rate start falling with the increased values of the nanoparticle concentration, the system is helped to stabilise.

Table 10.1 Effect of growth rate with its critical values 'n_c' for taking variations/modifications in one parameter at once

Variations/modifications in one parameter				*Effect on configuration*
Surface tension (*T*)	0.001	0.09	0.1	Highly stabilising
n_c	18.7	17.3	16.38	
Magnetic field (*H*)	1	1.5	2	Moderately stabilising
n_c	18.73	18.39	17.91	
Nanoparticles (ϕ)	0.001	0.009	0.05	Slowly stabilising
n_c	18.68	18.28	18.15	

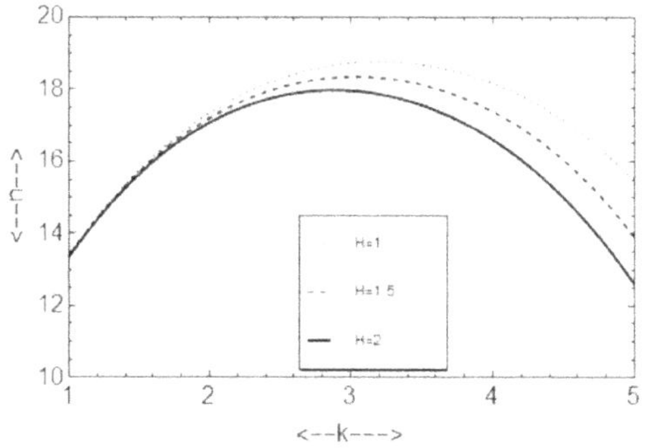

Figure 10.2 Effects of varying values of magnetic field on '*n*' growth rate under RTI.

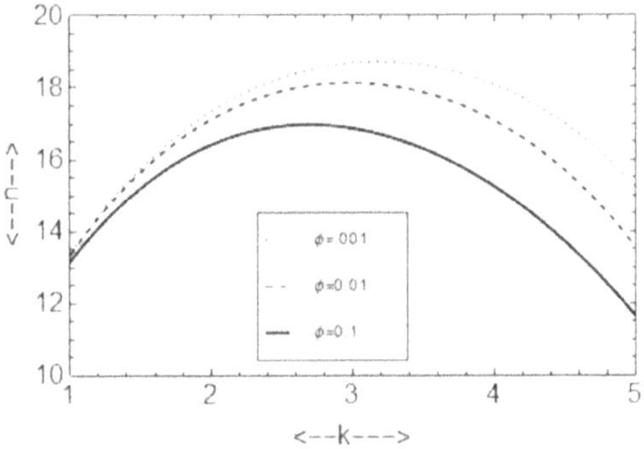

Figure 10.3 Effects of varying values of nanoparticles' volume fraction on '*n*' growth rate under RTI.

10.6.2 Variations/modifications considered in two parameter at once

For examining the behaviour of two parameters simultaneously, variations in those two parameters are taken by keeping other parameters as constant. Variations in two parameters are considered in the following aspects:

a. Both the parameters are increasing
b. One parameter is increasing while the other is decreasing

Thus, variations in two parameters are considered as (T, H), (ϕ, T) and (ϕ, H) at a time and the results are reported here.

10.6.2.1 Variation in (T,H) together under consideration

Let us consider the variation in a magnetic field along with surface tension keeping the volume fraction of nanoparticles as constant. In Figure 10.4, both (T, H) are taken in increasing order with values as (0.1, 1), (0.2, 1.5) and (0.3, 2). It is observed that the combined effect of these two variables has put the unstable system into fast stabilising mode with their proportionate increased values. It is stated that as the magnetic field and surface tension are increased, the critical values of growth rate meet a great fall and set the system to a rapid stabilisation.

Now if the values of 'T' are to be considered in increasing order while the values of the magnetic field are taken in decreasing order like (0.1, 2), (0.2, 1.5) and (0.3, 1) as in Figure10.5, then the system experiences a quick stabilisation for the unstable mode and RTI is suppressed. In Table 10.2, this observation is clearly depicted with the critical values 'n_c' of growth rate which is continuously declining for increasing the values of surface tension and decreasing the values of the magnetic field. It is observed that in Figure 10.5, the rate of stabilisation is higher than that of Figure10.4.

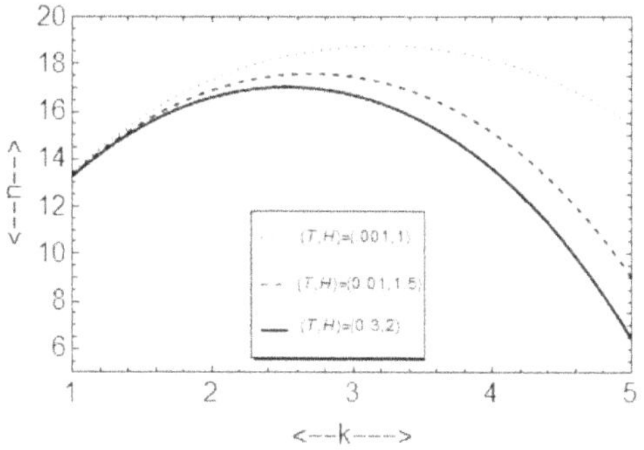

Figure 10.4 Effects of increasing values of (T, H) keeping ϕ constant 'n' growth rate under RTI.

Table 10.2 Effect of growth rate with its critical values 'n_c' for taking variations/modifications in two parameters at once

Variations/modifications in two parameters at once					*Remarks*
Both variables are inc	(H,T)	(1,0.1)	(1.5, 0.3)	(2, 0.5)	Stabilising rapidly
	n_c	18.73	17.47	16.638	
	(ϕ,T)	(0.009, .1)	(0.05, 0.2)	(0.09, 0.3)	Destabilising
	n_c	18.41	18.49	18.563	
	(ϕ,H)	(0.009, 1)	(0.05, 1.5)	(0.09, 2)	Destabilising quickly
	n_c	17.99	18.58	18.95	
Ist variable is increasing and IInd variable is decreasing	(T, H)	(0.1, 2)	(0.3, 1.5)	(0.5, 2)	Stabilising quickly
	n_c	18.84	17.47	16.54	
	(ϕ,T)	(0.009, .3)	(0.05, 0.2)	(0.09, 0.1)	Highly destabilising
	n_c	17.59	18.49	19.29	
	(ϕ,H)	(0.009, 2)	(0.05, 1.5)	(0.09, 1)	Destabilising
	n_c	18.09	18.53	18.91	

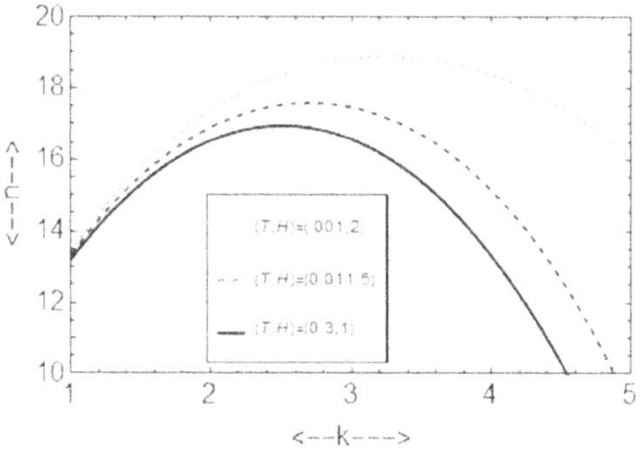

Figure 10.5 Effects of increasing values of T and decreasing values of H keeping ϕ constant on 'n' growth rate under RTI.

10.6.2.2 Variation in (ϕ, T) together under consideration

In this case, variations in the volume fraction of nanoparticles and surface tension (ϕ, T) parameters are considered and magnetic field H is to be taken as constant. First, let us consider the values of (ϕ, T) in increasing order as (0.009, 0.1), (0.05, 0.2) and (0.09, 0.3). In Figure 10.6, the behaviour of curves is clearly inclined toward destabilisation, but the destabilisation is quite slow as there is a slight increase in the critical values 'n_c' of the growth rate. But if the value of 'ϕ' is being increased along with a decrease in the values of 'T', the system has ensured a sudden rise in the critical values 'n_c' of growth rate. Thus, in Figure 10.7, the curves exhibit rapid destabilisation as compared with Figure 10.6.

In Table 10.2, critical values 'n_c' of growth rate for the combined effect of increased values of surface tension and the volume fraction of nanoparticles are shown, and it is observed that the system seems to infuse the onset of RTI but at a very moderate rate. Moreover, the system exhibits the high destabilisation for RTI whenever the values of surface tension are decreased

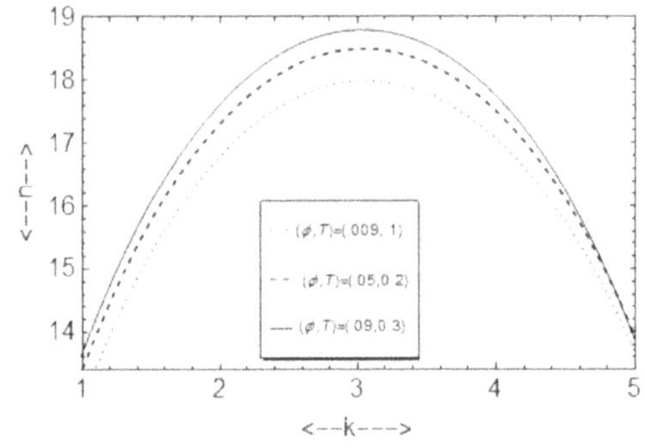

Figure 10.6 Effects of increasing values of (ϕ, T) keeping H constant on 'n' growth rate under RTI.

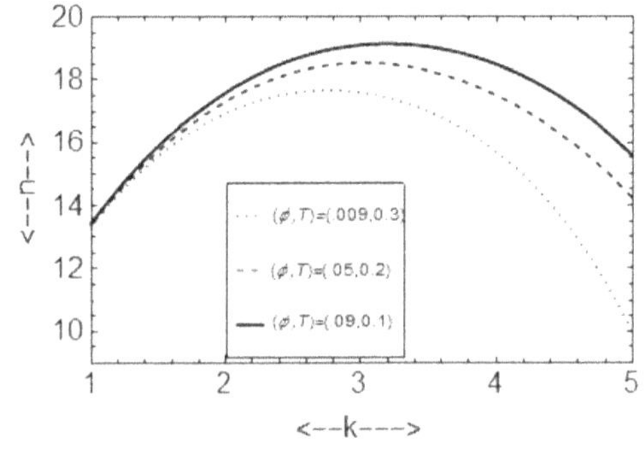

Figure 10.7 Effects of increasing values of ϕ and decreasing T keeping H constant on 'n' growth rate under RTI.

and the concentration of nanoparticles is increased. Thus, this combination leads to a significant impact on the perturbed system and encourages accelerating the instability.

10.6.2.3 Variation in (ϕ, H) together under consideration

In this part, the volume fraction of nanoparticles and magnetic fields are considered varying parameters whereas surface tension is taken as constant. In Figure 10.8, the values of (ϕ, H) are considered in increasing order as (0.009, 1), (0.05, 1.5) and (0.09, 2). It is depicted that for the perturbed system, the increased values of (ϕ, H) are contributing to hasten the destabilisation at a very good pace. Figure 10.9, with combination of increasing values of 'ϕ 'and decreasing values of 'H', is also stimulating destabilisation. For better insights, critical values of growth rate parameter for different combinations are calculated and put in Table 10.2. In Table 10.2, we may numerically confirm that the combined effect of two varying parameters is much stronger.

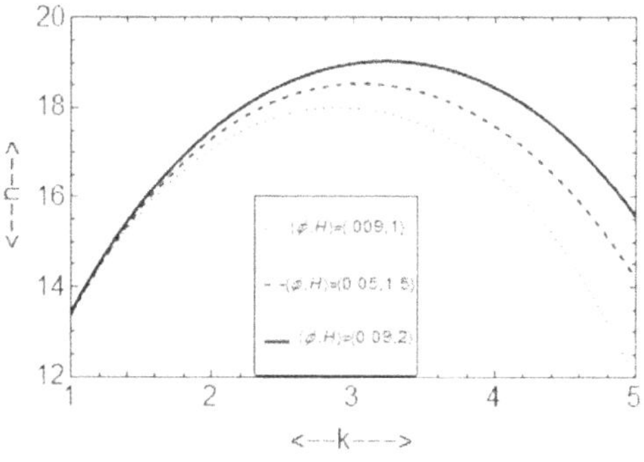

Figure 10.8 Effects of increasing values of (ϕ, H) keeping T constant on 'n' growth rate under RTI.

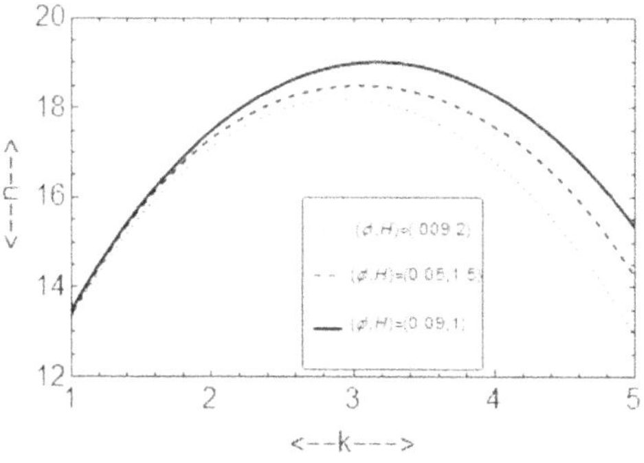

Figure 10.9 Effects of increasing values of ϕ and decreasing H keeping T constant on 'n' growth rate under RTI.

10.6.3 Variations/modifications considered in all three parameter at once

In this section, for examining the behaviour of three varying parameters taken together at a time, the following cases are considered:

a. All the parameters are varying in increasing order,
b. Two parameters are increasing while the other is decreasing,
c. One parameter is increasing while the other two are decreasing.

Let us consider that all values of (ϕ, H, T) parameters are in increasing order as (0.009, 1, 0.1), (0.05, 1.5, 0.2) and (0.09, 2, 0.3) and the results are presented in Figure 10.10. It portrays the fact that if these three parameters are increasing together then the system accelerates the destabilising process as critical values 'n_c' of the growth rate are also hiked with the increased values of these three parameters together. Figure 10.11 shows the results when the values of two parameters and H vary in increasing order and the values of 'T' are taken in decreasing order. The behaviour of curves is observed as highly destabilising. It is also compared that there is a significant rise in the critical values 'n_c' of the growth rate in Figure 10.11 as compared to Figure 10.10. This highly destabilisation quickens the onset of RTI in the configuration.

Thus, it is noted that the combined effect of these three varying parameters is much more superior to that of the individual parameter under consideration. Now if only 'ϕ' is taken with increased values while 'H' and 'T' are with decreased values then the combined effect of these values sets the system in the highly destabilising mode which helps to outset RTI in the perturbed configuration. This effect is presented in Figure 10.12.

As per Table 10.3, through the various critical values with the combination of magnetic field, surface tension and volume fraction of nanoparticles, it is established that the rate of destabilisation is very different

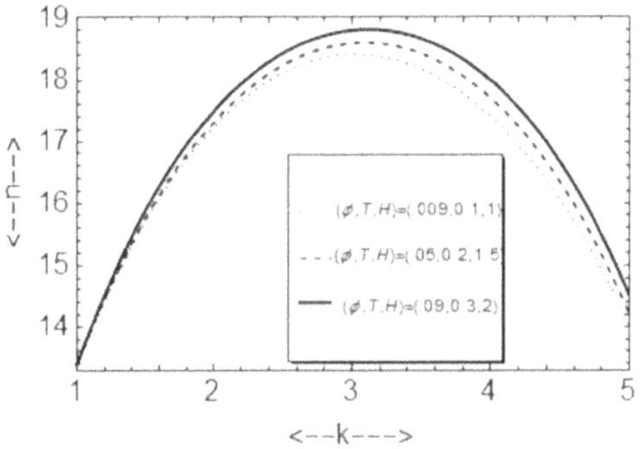

Figure 10.10 Effects of increasing values of (ϕ, H, T) on 'n' growth rate under RTI.

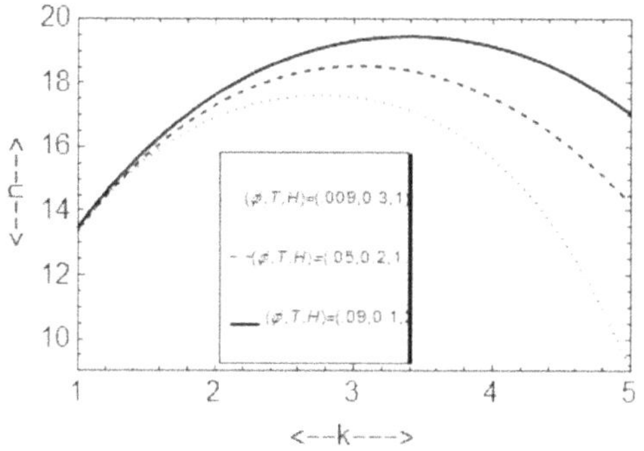

Figure 10.11 Effects of increasing values of (ϕ, H) but decreasing values of T on 'n' growth rate under RTI.

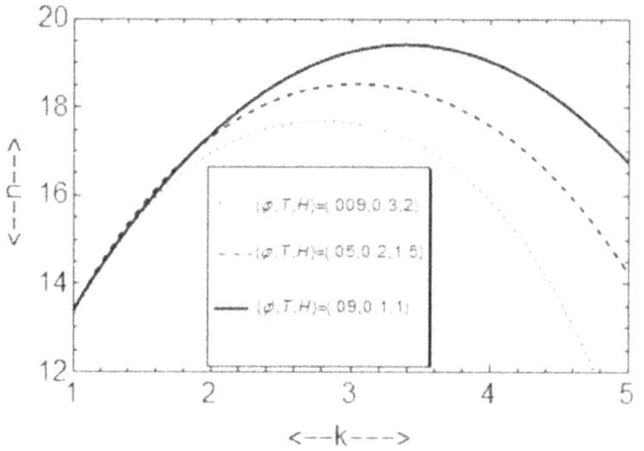

Figure 10.12 Effects of increasing values of ϕ but decreasing values of (H, T) on 'n' growth rate under RTI.

Table 10.3 Effect of growth rate with its critical values 'n_c' for taking variations/modifications in all parameters at once

Variations/modifications in all parameters with their inc. and dec. order					*Remarks*
I, II, III inc.	(ϕ,H,T)	(0.009, 1,0.1)	(0.05,1.5,0.2)	(0.09, 2, 0.3)	Slightly destabilising
	n_c	18.41	18.53	18.638	
Two variables inc. and IIIrd variable dec.	(ϕ,H,T)	(0.009, 1,0.3)	(0.05,1.5,0.2)	(0.09, 2, 0.1)	Highly destabilising
	n_c	17.56	18.537	19.334	
Ist variable inc. two variables dec.	(ϕ,H,T)	(0.009, 2,0.3)	(0.05,1.5,0.2)	(0.09, 1, 0.1)	Highly destabilising
	n_c	17.65	18.532	19.261	

with the combination of increased values of ϕ along with increasing and decreasing values of other varying parameters. Therefore, the increased value of 'ϕ' seems an agent of accelerating the process of destabilisation in the configuration under the influence of RTI when the magnetic field and surface tension have increasing/decreasing values. The impact of varying all three parameters simultaneously is inclined to set the system more agreeable for RTI only while keeping the concentration of nanoparticles increasing.

10.7 CONCLUSION

This article formulates the RTI in nanofluids imposed on one another in the presence of magnetic field. The system is subjected for perturbation and these perturbation equations are analysed using normal mode technique.

The impact of surface tension, magnetic field and volume fraction of nanoparticles is analysed by taking variation in one/two/three parameters simultaneously and presented in the form of graphs and tables. It is observed that the critical value of the growth rate parameter shows a substantial increase or decrease while taking variation in two/three parameters simultaneously as compared to the single parameter variation. Thus, the stability of the system gets more influenced while varying two parameters simultaneously as compared to the variation in a single parameter. This stabilisation/destabilisation impact gets stronger when variation in all three parameters is considered.

Here the conclusion is parameter-wise:

a. **Surface Tension:** The system's stability is significantly impacted by changes in surface tension. The system becomes more stable as surface tension rises, resulting in a lower critical value of growth rate. In the context of the RTI in nanofluid layers, this suggests that increased surface tension acts as a stabilising parameter, helping to resist the beginning of instability.
b. **Magnetic Field:** The existence of a magnetic field also has a significant impact on the stability of the system. The system becomes more stable as the magnetic field is increased, which leads to a lower critical value. The RTI is thereby suppressed in two stacked nanofluid layers as a result of the magnetic field's regulating effect.
c. **Nanoparticle Volume Fraction:** This factor directly influences the stability of the nanofluid layers. A lower critical value of growth rate results from the system quickly stabilising as the volume fraction of nanoparticles rises. This suggests that a higher volume proportion of nanoparticles aids in system stabilisation and delays the beginning of instability.

d. **Combined Effects:** Specific tendencies become apparent when taking into account the combined influence of two parameters at once:
   - Rapid stabilisation is caused by an increase in surface tension and magnetic field and results in a decrease in the critical value.
   - Moderate stabilisation occurs when surface tension is reduced while the magnetic field is increased, which results in a lower critical value.
   - The system becomes unstable as the surface tension and nanoparticle volume fraction grow while the magnetic field remains constant.
   - High destabilisation results from decreasing surface tension and the volume percentage of nanoparticles while maintaining a steady magnetic field.
   - High instability is produced by increasing the magnetic field and the volume fraction of nanoparticles while maintaining a steady surface tension.
   - High instability results from decreasing the magnetic field while increasing the volume percentage of nanoparticles and maintaining a steady surface tension.

e. **Combined Effects of All Parameters:**
   - The system becomes unstable when the surface tension, magnetic field and volume fraction all increase at the same time.
   - The system is severely destabilised when all three parameters decrease simultaneously.
   - The system becomes significantly more unstable when surface tension and the magnetic field are reduced while the volume percentage of nanoparticles is increased.

In conclusion, the study of individual and combination parameter effects yields important knowledge about the intricate behaviour of the RTI in nanofluid layers. It emphasises the vital importance of surface tension, magnetic field and nanoparticle volume fraction in determining system stability and provides essential guidance for optimising nanofluid-based systems for a variety of practical applications. To advance the design and control of nanofluid technologies in fields such as microfluidics, nanotechnology and energy systems, it is essential to comprehend these parameter-wise influences.

Therefore, Table 10.4 is constructed to summarise the results of the study.

10.8 SUMMARY OF RESULTS

Variable parameter	*Figures*	*Surface tension*	*Magnetic field*	*Volume fraction of nanoparticle*	*Critical value of n_c*	*Effects of varying parameter in existence of nanoparticles*
One	Figures 10.1–10.3	Figure 10.1	Figure10.2	Figure 10.3	↓	Stabilising
Variation in two parameters at a time	Figure 10.4	(Inc.)	(Inc.)	'c'	↓	Stabilising rapidly
	Figure 10.5	(Dec.)	(Inc.)	'c'	↓	Finely stabilising
	Figure 10.6	(Inc.)	'c'	(Inc.)	↑	Destabilising
	Figure 10.7	(Dec.)	'c'	(Inc.)	↑	Destabilising highly
	Figure 10.8	'c'	(Inc.)	(Inc.)	↑	Destabilisation quickly
	Figure 10.9	'c'	(Dec.)	(Inc.)	↑	Highly unstable
Variation in three parameters at a time	Figure 10.10	(Inc.)	(Inc.)	(Inc.)	↑	Destabilising
	Figure 10.11	(Dec.)	(Inc.)	(Inc.)	↑	Destabilising at higher pace
	Figure 10.12	(Dec.)	(Dec.)	(Inc.)	↑	Highly unstable

Note: In Table 10.4, 'c' refers to a constant.

NOMENCLATURE

$\delta(z - z_s)$	Dirac delta function
μ	viscosity of nanofluid in both of layers
μ_1	viscosity of nanofluid in lower layer
μ_2	viscosity of nanofluid in upper layer
μ_e	magnetic permeability
ρ	total density of the nanofluid in configuration
ρ_2	total density of the nanofluid in upper layer
ρ_f	total density of base fluid
ρ_{f1}	density of base fluid in lower layer
ρ_{f2}	density of base fluid in upper layer
ρ_p	total density of the nanoparticles ρ_1
ρ_{p1}	density of nanoparticles in lower layer
ρ_{p2}	density of nanoparticles in upper layer
ϕ	Volume fraction of nanoparticles in both of layers
ϕ_1	Volume fraction of nanoparticles in lower layer
ϕ_2	Volume fraction of nanoparticles in upper layer
D_B	the Brownian diffusion coefficient
h	components of magnetic field in all directions
H	vertical Magnetic field
p	pressure
T	Surface tension
v	velocity of fluid

REFERENCES

Ahuja, J. & Girotra, P. (2021a). Analytical and numerical investigation of Rayleigh-Taylor instability in nanofluids. *Pramana-Journal of Physics*, *95*, 25 (1–12).

Ahuja, J. & Girotra, P. (2021b). Rayleigh Taylor instability in nanofluids through porous medium. *Journal of Porous Media*, *24*, 49–70.

Ahuja, J. & Gupta, U. (2019). Rayleigh-Bénard convection for nanofluids for more realistic boundary conditions (rigid free and rigid-rgid) using Darcy model. *International Journal of Mathematical, Engineering and Management Sciences*, *4*, 139–156.

Ariel, P.D. & Aggarwala, B.D. (1979). The effect of a general oblique magnetic field on Rayleigh-Taylor insability. *Canadian Journal of Physics*, *57*, 1094–1102.

Bhadauria, B.S. (2007). Double diffusive convection in porous medium with modulated temperature on boundaries. *Transport in Porous Media*, *70*, 211.

Bhatia, P.K. (1974). Rayleigh-Taylor instability of viscous compressible plasma of variable density. *Astrophysics and Space Science*, *26*, 319–325.

Carlyle, J. & Hillier, A. (2017). The non-linear growth of magnetic Rayleigh-Taylor instability. *Astronomy and Astrophysics*, *605*, A101 (1–10).

Chakraborthy, B.B. (1979). A note on Rayleigh Taylor instability in presence of rotation. *Zeitschrift fur Angewandte Mathematik und Mechanik*, *59*, 651.

Chandrasekhar, S. ed. (1981). *Hydrodynamic and Hydromagnetic Stability*, Dover Publication, New York.

Choi, S.ed. (1995). Enhancing thermal conductivity of fluids with nanoparticles, in Developments Applications of Non-Newtonian Flows. ASME FED, New York. pp. 99–105.

Hide, R. (1956). The character of the equilibrium of a heavy, viscous, incompressible, rotating fluid of variable density II. Two special cases. *The Quarterly Journal of Mechanics and Applied Mathematics*, 9, 35.

Girotra, P., Ahuja, J. & Verma, D. (2021). Analysis of Rayleigh-Taylor instabilityin nanofluids with rotation. *Numerical Algebra, Control & Optimization*, *12*, 1–18.

Gupta, U., Ahuja, J. & Wanchoo, R.K. (2013). Magneto convection in nanofluid. *International Journal of Heat and Mass Transfer*, *64*, 1163–1171.

Nield, D.A. & Kuzenstov, A.V. (2014). Thermal instability in a porous medium layer saturated by a nanofluid: a revised model. *International Journal of Heat and Mass Transfer*, 68, 211–214.

Pacitto, G., Flament, C., Bacri, J.C. & Widom, M. (2000). Rayleigh-Taylor instability with magnetic fluids: experiment and theory. *Physical Review E*, 62, 7941–7947.

Rannacher, D. & Engel, A. (2007). Suppressing the Rayleigh-Taylor instability with a rotating magnetic field. *Physical Review E*, 75, 016311.

Rayleigh, L. (1900). *Investigation of the Character of the Equilibrium of an Incompressible Heavy Fluid of Variable Density*. Scientific Papers, J.W. Strutt, Ed., Cambridge University Press, Cambridge, II, pp.200–207.

Sengupta, A. & Joshi, B. (2023). Effects of stabilizing and destabilizing thermal gradients on reversed shear-stratified flows: combined Kelvin-Helmholtz Rayleigh-Taylor instability. *Physics of Fluids*, *35*(1), 012118.

Sengupta, A. & Verma, A.K. (2023). Role of unstable thermal stratifications on the Rayleigh-Taylor instability. *Computers & Fluids*, *252*, 105773. https://doi.org/10.1016/j.compfluid.2022.105773.

Segur, J.B. & Oberstar, H.E. (1951). Viscosity of glycerol and its aqueous solutions. *Industrial and Engineering Chemistry*, *43*, 2117–2120.

Sharma, R.C. & Bhardwaj, V.K. (1994). Rayleigh Taylor instability of Newtonian and Oldroydian viscoelastic fluids in porous medium. *Z. Naturforsch*, *Acta Physica Academiae Scientiarum Hungaricae*, *49*, 927–930.

Sharma, P.K. & Chhajlani, R.K. (1998). Effect of rotation on the Rayleigh Taylor instability of two superposed magnetized conducting plasma. *Physics of Plasama*, *5*, 2203–2209.

Sharma, R.C., Kumar, P. & Sharma, S. (2001). Rayleigh Taylor instability of Rivlin-Ericksen elastico-viscous fluid through porous medium. *Indian Journal of Physics*, *75B*, 337–340.

Sharma, P.K., Prajapati, R.P. & Chhajlani, R.K. (2010). Effect of surface tension and rotation on Rayleigh-Taylor instability of two superposed fluids with suspended particles. *Acta Physica Polonica A*, *118*, 576–584.

Shukla, A.K. & Awasthi, M.K. (2021). Rayleigh-Taylor instability with vertical magnetic field and heat transfer. *AIP Conference Proceedings*, *2352*, 020013(1–5).

Shukla, A.K., Awasthi, M.K. & Singh, S. (2023). Impact of heat and mass transport on Rayleigh-Taylor instability of Walter's B viscoelastic fluid layer. *Microgravity Science and Technology*, *35*(1), 3.

Taylor, G.I. (1950). The instability of liquid surfaces when accelerated in a direction perpendicular to their planes. *Proceedings of Royal Society London, A, 201*, 192–196.

Tzou, D.Y. (2008). Instability of nanofluids in natural convection. *ASME Journal of Heat Transfer*, *130*, 1–9.

Volk, A. & Khaler, C.J. (2018). Density model for aqueous glycerol solutions. *Experiments in Fluids*, *59*, 75 (1–4).

Chapter 11

Modelling of silicone oil-based Casson hybrid nanofluid across a porous rotating disk

Ashish Paul and Bhagyashri Patgiri
Cotton University

11.1 INTRODUCTION

Flow-induced by rotating disks are of great importance due to their various industrial applications such as medical equipment, rotating machinery, gas turbine rotor, viscometer, computer storage devices, and air cleaning machinery. Khan et al. [1] analysed Bingham–Papanastasiou flow due to an infinite rotating disk. Xun et al. [2] observed the heat Ostwald-de Waale fluid flow past a thickness-decreasing rotating disk. Rafiq et al. [3] experimented with unsteady fluid flow across a porous rotating with variable viscosity. Bhansali et al. [4] illustrated the impact of adding pin-fins on the heat transmission over a gas-turbine rotating disk's surface.

Hybrid nanofluids are one latest class of nanofluids attained by mixing two dissimilar nanoparticles. Due to the enriched thermophysical attributes of hybrid nanofluids, they have achieved humungous attention among researchers. Mohsan et al. [5] elaborated on convective heat transportation and flow features of Cu–Ag/water hybrid nanofluids. Sarkar et al. [6] reviewed recent research on hybrid nanofluids. Shafiq et al. [7] discussed utilisations of CNT-based hybrid nanofluid flow under Newtonian heating. Fallah et al. [8] explored heat transportation on SiC-TiO_2/DO hybrid nanofluid flow over one permeable spinning disk. Afterwards, several attempts were made by researchers on hybrid nanofluid flows over a rotatory disk – one may refer to the connected studies in References [9–18].

The non-Newtonian nature of fluids is encountered in the industries such as lubrication, mining, and biomedical. The flow analysis due to non-Newtonian fluid is of great industrial importance. Casson fluid is an important non-Newtonian fluid with tremendous importance in polymer and biomechanics industries. A decent number of studies have been conducted on Casson fluid flows over rotatory disks; one may observe the related studies in References [19–23].

The theory of porous media has been applied in several areas of science and engineering: soil mechanics, rock mechanics, biophysics, material science, construction, petroleum engineering, geoscience, etc. Thus, fluid flow via porous

 DOI: 10.1201/9781032712079-11

media has become a very significant subject of interest among researchers. A list of concerning literature can be observed in References [24–26].

MHD has recently earned noteworthy attention in many industries due to its usefulness in petroleum industries, medical industries (identifying diseases), cooling industries, nuclear reactors, etc. A list of research papers on MHD fluid flow due to rotatory disks can be observed in References [27–29].

By noticing the available concerning literature, one may conclude that no attention has been paid to study MHD flow on silicone oil–based Casson hybrid nanofluid past a rotatory disk. Silicone oils are widely used in many industrial utilisations, such as lens coatings, household product ingredients, glass vials, mechanical fluids, electrical insulating fluids, penetrating oil ingredients, and surface active agents. Thereby, our purpose here is to examine the MHD, 3D flow of Ag-Cu/silicone oil–based Casson hybrid nanofluid across one porous rotating disk with slip influences. The practical implementation of our work at engineering and industrial level is the manufacturing of silicon wafers, oil coolers, and many other appliances that are used in electrical and electronic industries.

11.2 MATHEMATICAL FORMULATION

MHD, incompressible, 3D mixed convective, steady flow of non-Newtonian (Casson) silicone oil–based hybrid nanofluid encountered by slip conditions are considered past a porous rotating disk. The physical framework of the hybrid nanofluid flow is sketched out in Figure 11.1. We adopt cylindrical coordinates (r, Φ, z). Moreover, (u, v, w) are taken as velocity components along (r, Φ, z) directions. Besides, T_W, T_∞ specify wall and ambient temperature, and notations C_W, C_∞ indicate wall and ambient concentration. Now, the equations which primarily govern this flow problem are (Waqas et al. [30] and Shafiq et al. [31]):

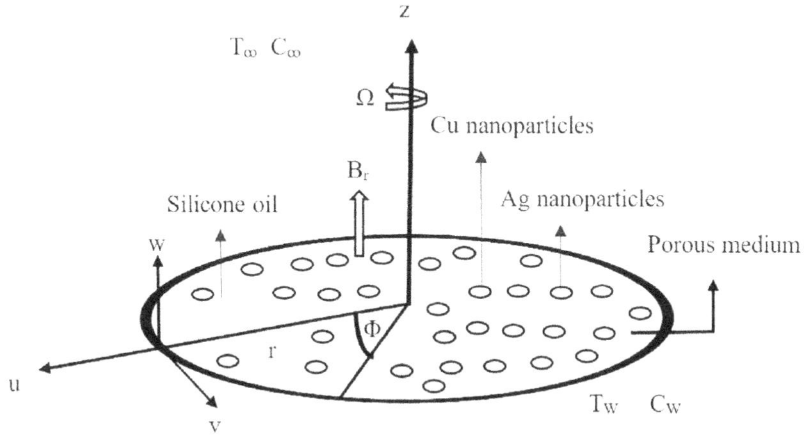

Figure 11.1 Flow sketch.

$$\frac{\partial u}{\partial r}+\frac{u}{r}+\frac{\partial w}{\partial z}=0 \tag{11.1}$$

$$\begin{aligned} u\frac{\partial u}{\partial r}+w\frac{\partial u}{\partial z}-\frac{v^2}{r}=&\frac{\mu_{hnf}}{\rho_{hnf}}\left(1+\frac{1}{\Xi}\right)\left(\frac{\partial^2 u}{\partial r^2}+\frac{1}{r}\frac{\partial u}{\partial r}-\frac{u}{r^2}+\frac{\partial^2 u}{\partial z^2}\right)\\ &-\frac{\mu_{hnf}}{\rho_{hnf}}\left(\frac{u}{L}\right)-\frac{\sigma_{hnf}}{\rho_{hnf}}uB_r^2+g\left[\gamma_T\left(T-T_\infty\right)\right] \end{aligned} \tag{11.2}$$

$$\begin{aligned} u\frac{\partial v}{\partial r}+w\frac{\partial v}{\partial z}-\frac{uv}{r}=&\frac{\mu_{hnf}}{\rho_{hnf}}\left(1+\frac{1}{\Xi}\right)\left(\frac{\partial^2 v}{\partial r^2}+\frac{1}{r}\frac{\partial v}{\partial r}-\frac{v}{r^2}+\frac{\partial^2 v}{\partial z^2}\right)\\ &-\frac{\mu_{hnf}}{\rho_{hnf}}\left(\frac{v}{L}\right)-\frac{\mu_{hnf}}{\rho_{hnf}}vB_r^2+g\left[\gamma_T\left(T-T_\infty\right)\right] \end{aligned} \tag{11.3}$$

$$u\frac{\partial w}{\partial r}+w\frac{\partial w}{\partial z}=\frac{\mu_{hnf}}{\rho_{hnf}}\left(1+\frac{1}{\Xi}\right)\left(\frac{\partial^2 w}{\partial r^2}+\frac{1}{r}\frac{\partial w}{\partial r}+\frac{\partial^2 w}{\partial z^2}\right)-\frac{\mu_{hnf}}{\rho_{hnf}}\left(\frac{w}{L}\right) \tag{11.4}$$

$$u\frac{\partial T}{\partial r}+w\frac{\partial T}{\partial z}=\frac{k_{hnf}}{\left(\rho c_p\right)_{hnf}}\left(\frac{\partial^2 T}{\partial r^2}+\frac{1}{r}\frac{\partial T}{\partial r}+\frac{\partial^2 T}{\partial z^2}\right) \tag{11.5}$$

$$u\frac{\partial C}{\partial r}+w\frac{\partial C}{\partial z}=\beta_{hnf}\left(\frac{\partial^2 C}{\partial z^2}\right) \tag{11.6}$$

And specific boundary assumptions are

$$\left\{\begin{array}{l} u=q\left(\dfrac{\partial u}{\partial z}\right),\quad v=r\Omega+q\left(\dfrac{\partial v}{\partial z}\right),\quad w=0,\quad T=T_W+q_t\left(\dfrac{\partial T}{\partial z}\right)\ \text{ and }\ C=C_w\ \text{ at }\ z=0\\ u\to 0,\quad v\to 0,\quad w\to 0,\quad T\to T_\infty\ \text{ and }\ C\to C_\infty\ \text{ as }\ z\to\infty \end{array}\right\} \tag{11.7}$$

Here, subscripts *hnf*, *nf* and *f* mean hybrid nanofluid, nanofluid and carrier fluid, respectively. Moreover, $\mu,\rho,k,\nu,c_p,\sigma,\Xi,\beta$ refer to dynamic viscosity, density, thermal conductivity, kinematic viscosity, heat capacity, electrical conductivity, Casson parameter, and mass diffusivity. Moreover, B_r,q,γ_T,L,q_t symbolise coefficients of magnetic field, velocity slip, thermal expansion, porosity, and thermal slip.

The physical features of Cu, Ag and silicone oil are presented in Table 11.1. Moreover, thermophysical correspondence connecting base fluid (silicone oil), nanofluid and hybrid nanofluid are stated as follows:

$$\mu_{nf} = \frac{\mu_f}{(1-\phi_1)^{2.5}}, \quad \mu_{bnf} = \frac{\mu_f}{(1-\phi_1)^{2.5}(1-\phi_2)^{2.5}},$$

$$\rho_{nf} = (1-\phi_1)\rho_f + \phi_1\rho_1, \quad \rho_{bnf} = (1-\phi_2)\left[(1-\phi_1)\rho_f + \phi_1\rho_1\right] + \phi_2\rho_2,$$

$$\left(\rho_{c_p}\right)_{nf} = (1-\phi_1)\left(\rho_{c_p}\right)_f + \phi_1\left(\rho_{c_p}\right)_1,$$

$$\left(\rho_{c_p}\right)_{bnf} = (1-\phi_2)\left[(1-\phi_1)\left(\rho_{c_p}\right)_f + \phi_1\left(\rho_{c_p}\right)_1\right] + \phi_2\left(\rho_{c_p}\right)_2,$$

$$k_{nf} = k_f\left\{\frac{k_1 + 2k_f - 2\phi_1(k_f - k_1)}{k_1 + 2k_f + \phi_1(k_f - k_1)}\right\}, \quad k_{bnf} = k_f\left\{\frac{k_2 + 2k_{nf} - 2\phi_2(k_{nf} - k_2)}{k_2 + 2k_{nf} + \phi_2(k_{nf} - k_{Ag})}\right\},$$

$$\sigma_{nf} = \sigma_f\left[\frac{\sigma_1 + 2\sigma_f - 2\phi_1(\sigma_f - \sigma_1)}{\sigma_1 + 2\sigma_f + \phi_1(\sigma_f - \sigma_1)}\right]$$

and $$\sigma_{bnf} = \sigma_f\left[\frac{\sigma_2 + 2\sigma_{nf} - 2\phi_2(\sigma_{nf} - \sigma_2)}{\sigma_2 + 2\sigma_{nf} + \phi_2(\sigma_{nf} - \sigma_2)}\right]$$

Here, ϕ_1, ϕ_2 mean volume fractions of Cu and Ag nanoparticles.

To change dimensional equations (11.1)–(11.7) into a dimensionless model, we assign the following similarity conversion (Rehman et al. [34]):

$$\left\{\begin{array}{c} \xi = \sqrt{\frac{2\Omega}{v_f}}\, z, \quad u = r\Omega m'(\xi), \quad v = r\Omega n(\xi), \quad w = -\sqrt{2\Omega\chi}\; m(\xi), \\ \Delta(\xi) = \frac{(T - T_\infty)}{(T_w - T_\infty)}, \quad \chi(\xi) = \frac{(C - C_\infty)}{(C_w - C_\infty)} \end{array}\right\} \tag{11.8}$$

Table 11.1 Physical features of silicone oil, Cu and Ag (Saranya et al. [32] and Divya et al. [33])

Name	*P (kg/m³)*	*c_p (J/kgK)*	*k (W/mK)*	*σ (S/m)*
Silicone oil	818	1966	0.1	1.5×10^{-4}
Cu	8933	385	401	5.96×10^{7}
Ag	10500	235	429	6.30×10^{7}

By employing equation (11.8) in equations (11.1)–(11.7), we perceive

$$m''' = \frac{P}{2}\cdot m' + \frac{V_1V_5}{\left(1+\frac{1}{\Xi}\right)}\cdot(M\cdot m') + \frac{V_1V_2}{2\left(1+\frac{1}{\Xi}\right)}\cdot\left[(m')^2 - 2m\cdot m'' - n^2 - N\cdot\Delta\right] \tag{11.9}$$

$$n'' = \frac{P}{2}\cdot n + \frac{V_1V_5}{\left(1+\frac{1}{\Xi}\right)}\cdot(M\cdot n) + \frac{V_1V_2}{2\left(1+\frac{1}{\Xi}\right)}\cdot\left[2m'\cdot n - 2m\cdot n' - N\cdot\Delta\right] \tag{11.10}$$

$$\Delta'' = \left(\frac{-V_3\cdot \Pr}{V_4}\right)\cdot m\cdot\Delta' \tag{11.11}$$

$$\chi'' = -m\cdot V_6\cdot Sc\cdot\chi' \tag{11.12}$$

and

$$\left\{\begin{matrix} m(0)=0,\ m'(0)=s\cdot m''(0),\ n(0)=1+s\cdot n'(0),\ \Delta(0)=1+s_t\cdot\Delta'(0)\text{ and }\Xi(0)=1\text{ at }\xi=0 \\ m(\infty)\to 0,\ m'(\infty)\to 0,\ n(\infty)\to 0,\ \Delta(\infty)\to 0\text{ and }\Xi(\infty)\to 0\text{ as }\xi\to\infty \end{matrix}\right\} \tag{11.13}$$

Here, prime mean differentiation w.r.t. ξ and

$$V_1 = \frac{\mu_f}{\mu_{hnf}},\ V_2 = \frac{\rho_{hnf}}{\rho_f},\ V_3 = \frac{(\rho_{c_p})_{hnf}}{(\rho_{c_p})_f},\ V_4 = \frac{k_{hnf}}{k_f},\ V_5 = \frac{\sigma_{hnf}}{\sigma_f},\ V_6 = \frac{\beta_f}{\beta_{hnf}}$$

Furthermore,

$$v_f = \frac{\mu_f}{\rho_f}\ (\text{Kinematic Viscosity}),\ Re = \frac{\Omega r^2}{v_f}(\text{Reynold's Number}),$$

$$M = \frac{\sigma_f B_r^2}{2\Omega\rho_f}\ (\text{Magnetic Parameter}),\ P = \frac{v_f}{\Omega L}(\text{ Porosity Parameter})$$

$$Gr = \frac{g r^2\gamma_T\ (T_W - T_\infty)}{(v_f)^2}(\text{Thermal Grasoff Number}),$$

$$N = \frac{Gr}{Re^2}(\text{ Mixed Convection Parameter}),$$

$$s = q\left(\sqrt{\frac{2\Omega}{\nu_f}}\right)(\text{Velocity slip parameter}),$$

$$Pr = \frac{\nu_f(\rho c_p)_f}{k_f}(\text{Prandtl number}),\ Sc = \frac{\nu_f}{D_f}(\text{Schimdt number}),$$

and $$s_t = q_t\left(\sqrt{\frac{2\Omega}{\nu_f}}\right)(\text{thermal slip parameter})$$

The dimensionless description of engineering quantities of curiosity are as follows:

Skin friction coefficient:

$$Cf_k Re^{0.5} = (1-\phi_1)^{-2.5}(1-\phi_2)^{-2.5}\left(1+\frac{1}{\Xi}\right)\sqrt{(m''(0))^2+(n'(0))^2}$$

Local Nusselt number:

$$Nu_k Re^{-0.5} = \left(\frac{-k_{hnf}}{k_f}\right)\cdot\Delta'(0) = -V_4\cdot\Delta'(0)$$

Local Sherwood number:

$$Sh_k Re^{-0.5} = -\chi'(0)$$

11.3 NUMERICAL METHOD AND VALIDATION

The coupled non-linear ODEs (11.9)–(11.12) with the boundary stipulations in (11.13) are handled numerically adopting the bvp4c scheme, a MATLAB-based finite difference tool. In this technique, we convert equations (11.9)–(11.13) into a series of first-order ODEs. The mesh size ξ_∞ and boundary layer thicknesses $\Delta\xi$ are perceived as 0.01 and 20.

Table 11.2 is formed to check the exactness of the used method with earlier communicated data in non-appearance of the Casson parameter and porous parameter. From Table 11.2, one may spot that an outstanding agreement is attained betwixt the results achieved by Mustafa et al. [35] and us.

Table 11.2 Comparison of $m''(0)$ for different M's values with Mustafa et al. [35]

M	s	$m''(0)$ (*Mustafa et al. [35]*)	$m''(0)$ (*Our output*)
0	0.25	0.259534	0.259534
0.4	0.25	0.146057	0.146057
0.8	0.25	0.096762	0.096762

11.4 RESULTS AND DISCUSSIONS

The main aspiration of this section is to discuss the impressions of several pertinent flow parameters on radial velocity $m'(\xi)$, tangential velocity $n(\xi)$, temperature $\Delta(\xi)$, and concentration $\chi(\xi)$ profiles and shearing stress, local Nusselt and Sherwood number.

Figures 11.2–11.5 are sketched out to check the effect of porosity parameter P on radial velocity $m'(\xi)$, tangential velocity $n(\xi)$, temperature $\Delta(\xi)$ and concentration $\chi(\xi)$ profiles. It is evident from this figures that P is an assisting factor for both $\Delta(\xi)$ and $\chi(\xi)$, whereas P reduces both $m'(\xi)$ and $n(\xi)$. Porosity provides resistance to hybrid nanofluid's motion and so both $m'(\xi)$ and $n(\xi)$ decrease as P increases. An increment in P leads to an increment in hybrid nanofluid's viscosity, resulting in increment in both $\Delta(\xi)$ and $\chi(\xi)$.

Figures 11.6–11.8 revealed that improved values of magnetic parameter elevate both $\Delta(\xi)$ and $\chi(\xi)$ and diminish radial velocity $m'(\xi)$. As M upsurges, the retarding body force also elevates, and thus $m'(\xi)$ decreases. Stronger values of M result in thicker thermal and concentration boundary layers, which eventually lead to amplification in both $\Delta(\xi)$ and $\chi(\xi)$.

Figures 11.9 and 11.10 are drawn to analyse the contribution of the Casson parameter Ξ on tangential velocity $n(\xi)$, and concentration $\Delta(\xi)$ profile. It may be deduced that both $n(\xi)$ and $\Delta(\xi)$ diminish for greater values of Ξ. Physically, for higher values of Ξ, the thicknesses of both momentum and thermal boundary layers shrink, and consequently, both $n(\xi)$ and $\Delta(\xi)$ decline.

Figure 11.11 portrays that the concentration profile $\chi(\xi)$ is decayed for amplifying values of Schmidt number Sc. For higher magnitudes of Sc, mass diffusivity of hybrid nanofluid depreciates, and consequently, concentration boundary layer thickness shrinks, and eventually, $\chi(\xi)$ also depletes.

Table 11.3 presents the contribution of Ag nanoparticle's volume fraction Φ_2, porosity parameter P, and Casson parameter Ξ on skin friction coefficient $Cf_k Re^{0.5}$, local Nusselt $(Nu_k Re^{-0.5})$ and Sherwood number $Sh_k Re^{-0.5}$. One may notice that $Cf_k Re^{0.5}$ escalates for larger values of Φ_2 and P, whereas depletes for larger magnitudes of Ξ. Moreover, heat

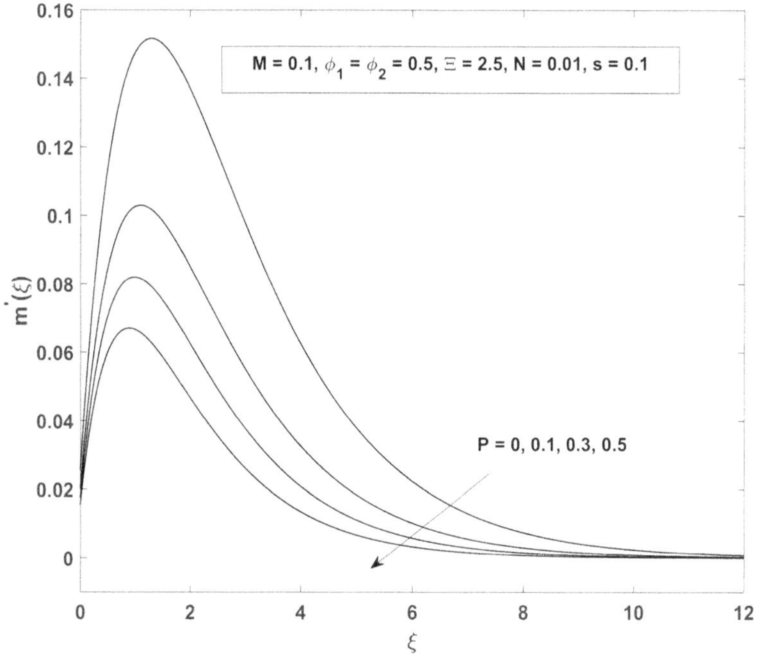

Figure 11.2 Impact of *P* on $m'(\xi)$.

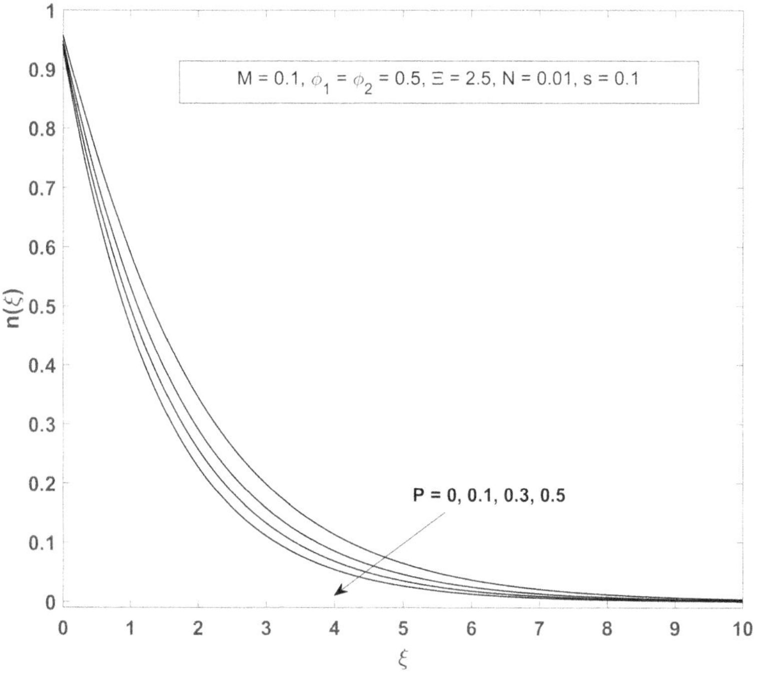

Figure 11.3 Impact of *P* on $n(\xi)$.

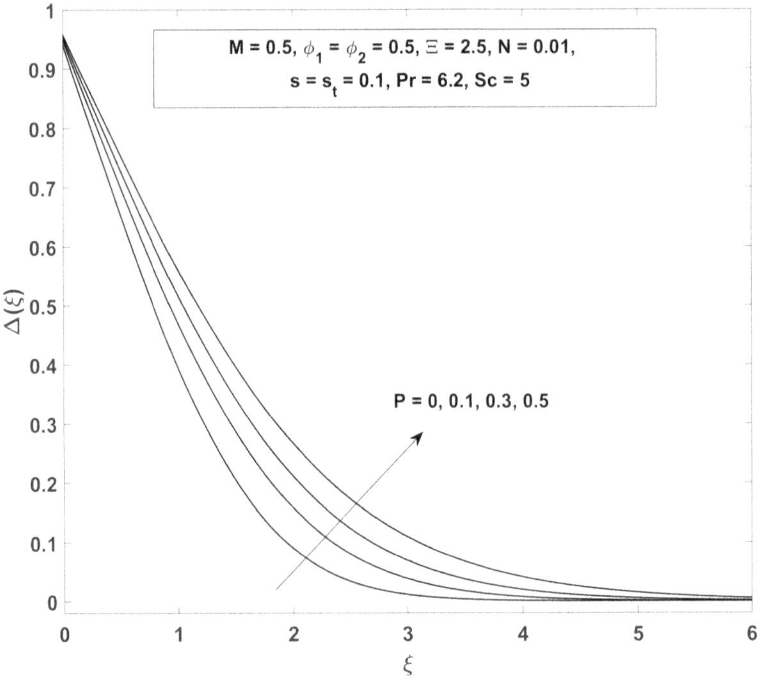

Figure 11.4 Impact of *P* on $\Delta(\xi)$.

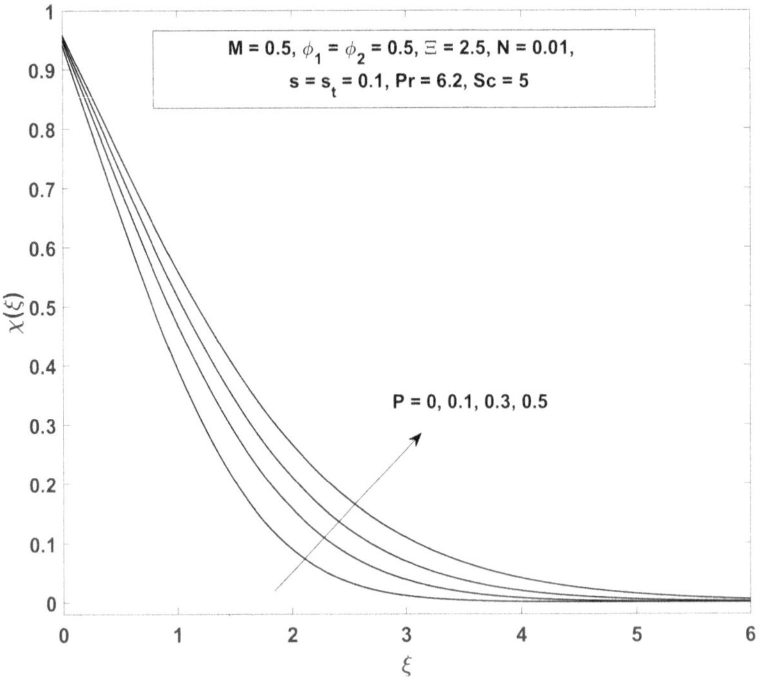

Figure 11.5 Impact of *P* on $\chi(\xi)$.

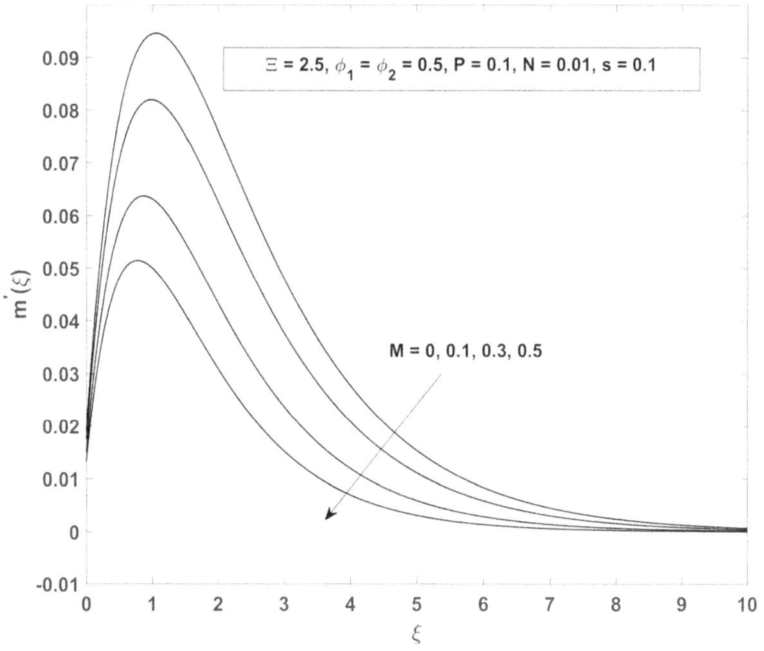

Figure 11.6 Impact of *M* on $m'(\xi)$.

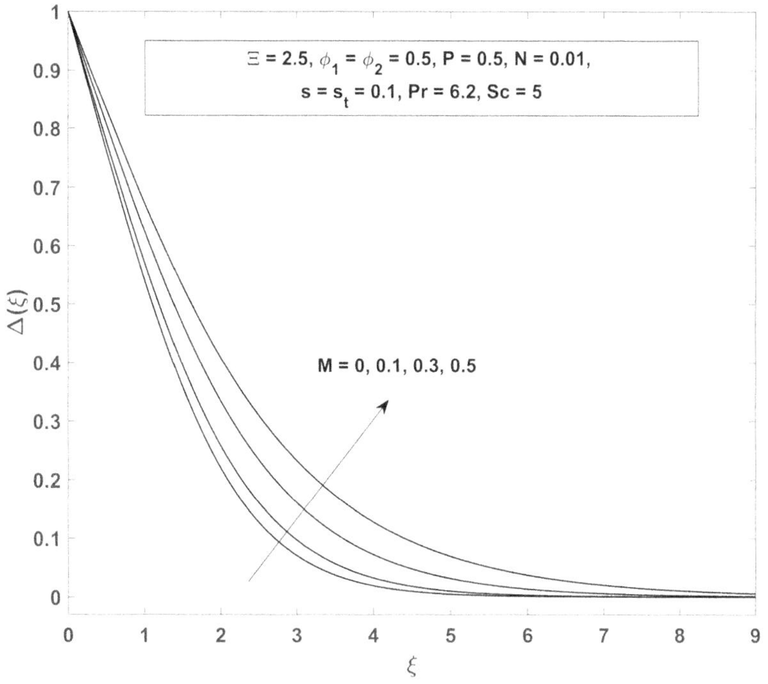

Figure 11.7 Impact of *M* on $\Delta(\xi)$.

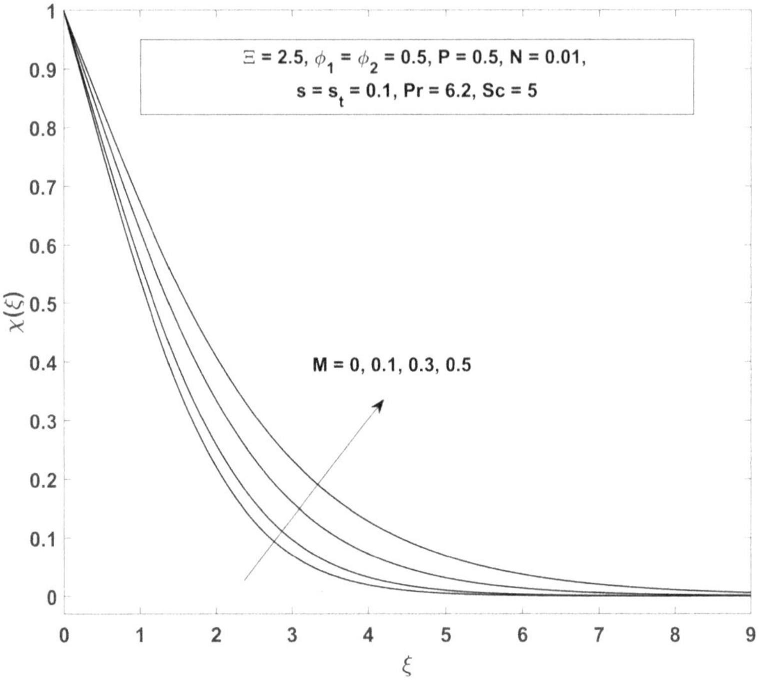

Figure 11.8 Impact of M on $\chi(\xi)$.

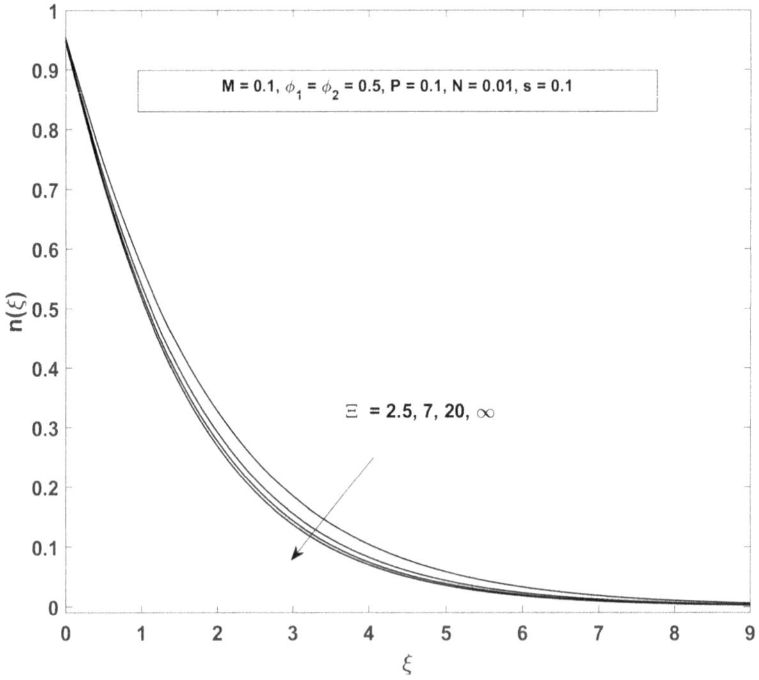

Figure 11.9 Impact of Ξ on $n(\xi)$.

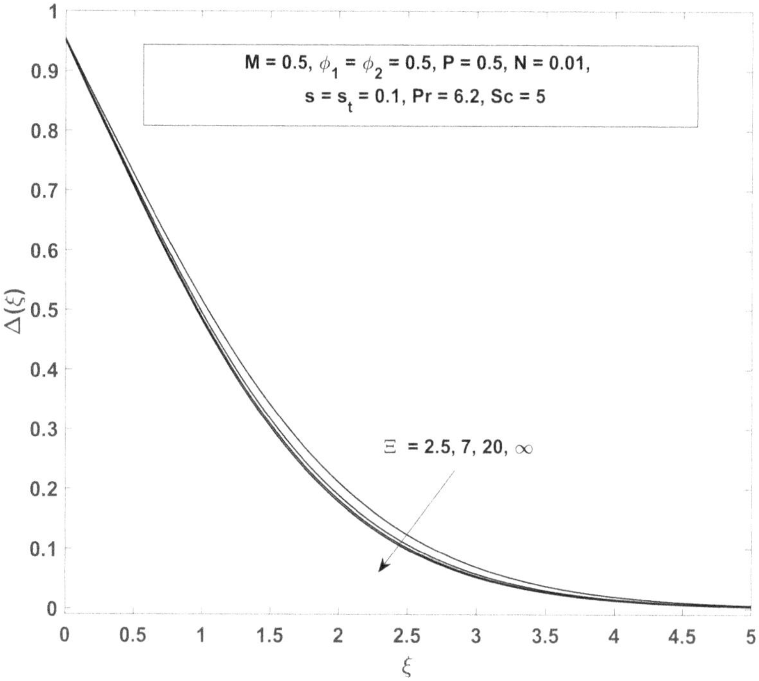

Figure 11.10 Impact of Ξ on $\Delta(\xi)$.

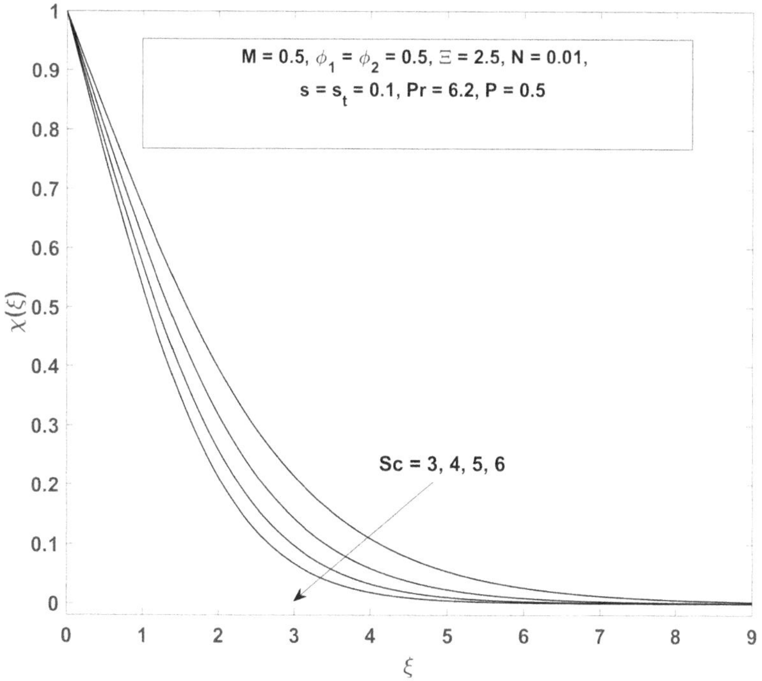

Figure 11.11 Impact of *Sc* on $\chi(\xi)$.

Table 11.3 Influence of embedded parameters on Cf_k, Nu_k, and Sh_k when $\phi_1 = 0.03$, $M = s = s_t = 0.1$, $Pr = 6.2$, $N = 0.01$, $Sc = 4$

Φ_2	P	Ξ	Cf_k	Nu_k	Sh_k
0	0.6	2.5	0.972969	0.469167	0.325435
0.01			0.991974	0.472037	0.321259
0.03			1.032362	0.477006	0.312695
	0.5		0.989944	0.501666	0.334139
	0.4		0.947730	0.527577	0.356800
	0.5	1.5	1.131236	0.440115	0.372702
		3.5	0.929948	0.468724	0.400449
		5	0.884417	0.475787	0.407256

transportation rate $(Nu_k Re^{-0.5})$ is amplified by ~1.7% in the case of hybrid nanofluid than nanofluid. Moreover, heat $(Nu_k Re^{-0.5})$ and mass $(Sh_k Re^{-0.5})$ transportation rates are depleted for higher values of P; moreover, Φ_2 acts as a diminishing factor for $Sh_k Re^{-0.5}$. Besides, Table 11.3 demonstrates that Casson parameter intensifies the heat and mass distribution rate in the silicone-based Casson hybrid nanofluid, respectively, by 8.16% and 9.27% and regresses the skin friction by 21.82%.

Thus, for the purpose of mitigating the advantage of skin friction as well as raising the advantages of heat and mass transpiration that we basically seek from an engineering point of view, the strength of the Casson parameter must be intensified, whereas the strength of the porous parameter should be minimised.

11.5 CONCLUSION

The major ambition of this exploration is to elaborate the attributes of heat and mass transpiration on Casson hybrid nanofluid mixed convective across a porous rotating disk under the attendance of slip conditions and magnetic field. We pick Cu and Ag as nanoparticles and silicone oil as the carrier fluid. The repercussions of noteworthy flow variables upon flow profiles are inspected. Besides, the heat and mass transpiration mechanisms are also exploited. For the numerical simulation of our problem, the bvp4c scheme is implemented. The significant results concluded from this inspection are as follows:

- An increment in porous parameter P hinders both velocity gradients, whereas it amplifies temperature and concentration dispersions.
- Magnetic parameter M is an assisting factor for both temperature and concentration dispersions, but it hinders radial velocity.

- The Casson parameter is a dwindling factor for tangential velocity and temperature profiles.
- Schmidt number is a hindering factor for concentration dispersion.
- The rise in Ag nanoparticles volume fraction allows the heat transpiration rate to augment by 1.7%.
- Casson parameter reduces skin friction by 21.82%.
- As the potency of the Casson parameter improves, both heat and mass transpiration rates upsurge separately by 8.16% and 9.27%.
- The porous parameter P raises the skin friction coefficient.
- The porous parameter P regresses the heat and mass transpiration rate.

NOMENCLATURE

Δ	Dimensionless temperature
Ω	Angular velocity
β	Mass diffusivity
γ_T	Coefficient of thermal expansion
μ	Dynamic viscosity
v_f	Kinematic viscosity
ξ	Dimensionless similarity variable
ρ	Density
σ	Electrical conductivity
χ	Dimensionless concentration
C	Concentration
C_∞	Free stream concentration
c_p	Specific heat capacity
C_w	Surface concentration
g	Acceleration due to gravity
Gr	Thermal Grashof number
k	Thermal conductivity
L	Coefficient of porosity
M	Magnetic parameter
m', n	Dimensionless velocity components
N	Mixed convection parameter
P	Porosity Parameter
Pr	Prandtl number
q	Coefficient of velocity slip
q_t	Coefficient of thermal slip
r, Φ, z	Cylindrical coordinates
Re	Reynold number
s	Velocity slip parameter
Sc	Schmidt number
T	Temperature

T_∞ Ambient temperature
T_w Surface temperature
u, v, w Radial, tangential and axial components of velocity

REFERENCES

[1] Khan, N. A., & Sultan, F. (2018). Numerical analysis for the Bingham-Papanastasiou fluid flow over a rotating disk. *Journal of Applied Mechanics and Technical Physics*, 59, 638–644. https://doi.org/10.1134/S0021894418040090.

[2] Xun, S., Zhao, J., Zheng, L., Chen, X., & Zhang, X. (2016). Flow and heat transfer of Ostwald-de Waele fluid over a variable thickness rotating disk with index decreasing. *International Journal of Heat and Mass Transfer*, 103, 1214–1224. https://doi.org/10.1016/j.ijheatmasstransfer.2016.08.066.

[3] Rafiq, T., Mustafa, M., & Farooq, M. A. (2020). Modelling heat transfer in fluid flow near a decelerating rotating disk with variable fluid properties. *International Communications in Heat and Mass Transfer*, 116, 104673. https://doi.org/10.1016/j.icheatmasstransfer.2020.104673.

[4] Bhansali, P. S., Ramakrishnan, K. R., & Ekkad, S. V. (2022). Effect of pin fins on jet impingement heat transfer over a rotating disk. *Journal of Heat Transfer*, 144(4), 042303. https://doi.org/10.1115/1.4053371.

[5] Mohsan, H., Marin, M., Ellahi, R., & Alamri, S. Z. (2018). Exploration of convective heat transfer and flow characteristics synthesis by Cu-Ag/water hybrid-nanofluids. *Heat Transfer Research*, 49, 1837–1848.

[6] Sarkar, J., Ghosh, P., & Adil, A. (2015). A review on hybrid nanofluids: recent research, development and applications. *Renewable and Sustainable Energy Reviews*, 43, 164–177.

[7] Shafiq, A., & Nadeem, S. (2020). Application of CNT-based micropolar hybrid nanofluid flow in the presence of Newtonian heating. *Applied Nanoscience*, 10, 1–13.

[8] Fallah, B., Dinarvand, S., EftekhariYazdi, M., Rostami, M. N., & Pop, I. (2019). MHD flow and heat transfer of SiC-TiO_2/DO hybrid nanofluid due to a permeable spinning disk by a novel algorithm. *Journal of Applied and Computational Mechanics*, 5(5), 976–988. https://doi.org/10.22055/JACM.2019.27997.1449.

[9] Tassaddiq, A., Khan, S., Bilal, M., Gul, T., Mukhtar, S., Shah, Z., &Bonyah, E. (2020). Heat and mass transfer together with hybrid nanofluid flow over a rotating disk. *AIP Advances*, 10(5). https://doi.org/10.1063/5.0010181.

[10] Jayadevamurthy, P. G. R., Rangaswamy, N. K., Prasannakumara, B. C., & Nisar, K. S. (2020). Emphasis on unsteady dynamics of bioconvective hybrid nanofluid flow over an upward-downward moving rotating disk. *Numerical Methods for Partial Differential Equations*. https://doi.org/10.1002/num.22680.

[11] Gul, T., Kashifullah, Bilal, M., Alghamdi, W., Asjad, M. I., &Abdeljawad, T. (2021). Hybrid nanofluid flow within the conical gap between the cone and the surface of a rotating disk. *Scientific Reports*, 11(1), 1180. https://doi.org/10.1038/s41598-020-80750-y.

[12] Reddy, M. G., Kumar, N., Prasannakumara, B. C., Rudraswamy, N. G., & Kumar, K. G. (2021). Magnetohydrodynamic flow and heat transfer of a hybrid nanofluid over a rotating disk by considering Arrhenius energy. *Communications in Theoretical Physics*, 73(4), 045002. https://doi.org/10.1088/1572-9494/abdaa5.
[13] Waqas, H., Farooq, U., Naseem, R., Hussain, S., & Alghamdi, M. (2021). Impact of MHD radiative flow of hybrid nanofluid over a rotating disk. *Case Studies in Thermal Engineering*, 26, 101015. https://doi.org/10.1016/j.csite.2021.101015.
[14] Shoaib, M., Raja, M. A. Z., Sabir, M. T., Nisar, K. S., Jamshed, W., Felemban, B. F., & Yahia, I. S. (2021). MHD hybrid nanofluid flow due to rotating disk with heat absorption and thermal slip effects: an application of intelligent computing. *Coatings*, 11(12), 1554. https://doi.org/10.3390/coatings11121554.
[15] Naveed Khan, M., Ahmad, S., Ahammad, N. A., Alqahtani, T., & Algarni, S. (2022). Numerical investigation of hybrid nanofluid with gyrotactic microorganism and multiple slip conditions through a porous rotating disk. *Waves in Random and Complex Media*, 1–16. https://doi.org/10.1080/17455030.2022.2055205.
[16] Kumar, M., & Mondal, P. K. (2022). Irreversibility analysis of hybrid nanofluid flow over a rotating disk: effect of thermal radiation and magnetic field. *Colloids and Surfaces A: Physicochemical and Engineering Aspects*, 635, 128077. https://doi.org/10.1016/j.colsurfa.2021.128077.
[17] Waini, I., Ishak, A., & Pop, I. (2022). Multiple solutions of the unsteady hybrid nanofluid flow over a rotating disk with stability analysis. *European Journal of Mechanics-B/Fluids*, 94, 121–127. https://doi.org/10.1016/j.euromechflu.2022.02.011.
[18] Vijay, N., & Sharma, K. (2023). Magnetohydrodynamic hybrid nanofluid flow over a decelerating rotating disk with Soret and Dufour effects. *Multidiscipline Modeling in Materials and Structures*, 19(2), 253–276. https://doi.org/10.1108/MMMS-08-2022-0160.
[19] Saeed, A., Shah, Z., Islam, S., Jawad, M., Ullah, A., Gul, T., & Kumam, P. (2019). Three-dimensional Casson nanofluid thin film flow over an inclined rotating disk with the impact of heat generation/consumption and thermal radiation. *Coatings*, 9(4), 248. https://doi.org/10.3390/coatings9040248.
[20] Waqas, H., Naseem, R., Muhammad, T., & Farooq, U. (2021). Bioconvection flow of Casson nanofluid by rotating disk with motile microorganisms. *Journal of Materials Research and Technology*, 13, 2392–2407. https://doi.org/10.1016/j.jmrt.2021.05.092.
[21] Shehzad, S. A., Mabood, F., Rauf, A., Izadi, M., & Abbasi, F. M. (2021). Rheological features of non-Newtonian nanofluids flows induced by stretchable rotating disk. *PhysicaScripta*, 96(3), 035210. https://doi.org/10.1088/1402-4896/abd652.
[22] Naveen Kumar, R., Gowda, R. P., Gireesha, B. J., & Prasannakumara, B. C. (2021). Non-Newtonian hybrid nanofluid flow over vertically upward/downward moving rotating disk in a Darcy-Forchheimer porous medium. *The European Physical Journal Special Topics*, 230, 1227–1237. https://doi.org/10.1140/epjs/s11734-021-00054-8.

[23] Gowda, R. P., Rauf, A., Naveen Kumar, R., Prasannakumara, B. C., & Shehzad, S. A. (2022). Slip flow of Casson-Maxwell nanofluid confined through stretchable disks. *Indian Journal of Physics*, 96(7), 2041–2049. https://doi.org/10.1007/s12648-021-02153-7.

[24] Sharma, K., Vijay, N., Makinde, O. D., Bhardwaj, S. B., Singh, R. M., & Mabood, F. (2021). Boundary layer flow with forced convective heat transfer and viscous dissipation past a porous rotating disk. *Chaos, Solitons & Fractals*, 148, 111055. https://doi.org/10.1016/j.chaos.2021.111055.

[25] Zhou, S. S., Bilal, M., Khan, M. A., & Muhammad, T. (2021). Numerical analysis of thermal radiative maxwell nanofluid flow over-stretching porous rotating disk. *Micromachines*, 12(5), 540. https://doi.org/10.3390/mi12050540.

[26] Upadhya, S. M., Devi, R. R., Raju, C. S. K., & Ali, H. M. (2021). Magnetohydrodynamic nonlinear thermal convection nanofluid flow over a radiated porous rotating disk with internal heating. *Journal of Thermal Analysis and Calorimetry*, 143, 1973–1984. https://doi.org/10.1007/s10973-020-09669-w.

[27] Turkyilmazoglu, M. (2022). Flow and heat over a rotating disk subject to a uniform horizontal magnetic field. *ZeitschriftfürNaturforschung A*, 77(4), 329–337. https://doi.org/10.1515/zna-2021-0350.

[28] Hosseinzadeh, K., Mogharrebi, A. R., Asadi, A., Sheikhshahrokhdehkordi, M., Mousavisani, S., & Ganji, D. D. (2022). Entropy generation analysis of mixture nanofluid ($H_2O/c_2H_6O_2$)-Fe_3O_4 flow between two stretching rotating disks under the effect of MHD and nonlinear thermal radiation. *International Journal of Ambient Energy*, 43(1), 1045–1057. https://doi.org/10.1080/01430750.2019.1681294.

[29] Bilal, M., Ayed, H., Saeed, A., Brahmia, A., Gul, T., & Kumam, P. (2022). The parametric computation of nonlinear convection magnetohydrodynamic nanofluid flow with internal heating across a fixed and spinning disk. *Waves in Random and Complex Media*, 1–16. https://doi.org/10.1080/17455030.2022.2042621.

[30] Waqas, H., Shehzad, S. A., Khan, S. U., & Imran, M. (2019). Novel numerical computations on flow of nanoparticles in porous rotating disk with multiple slip effects and microorganisms. *Journal of Nanofluids*, 8(7), 1423–1432. https://doi.org/10.1166/jon.2019.1702.

[31] Shafiq, A., Rasool, G., Alotaibi, H., Aljohani, H. M., Wakif, A., Khan, I., & Akram, S. (2021). Thermally enhanced Darcy-Forchheimer Casson-water/glycerine rotating nanofluid flow with uniform magnetic field. *Micromachines*, 12(6), 605. https://doi.org/10.3390/mi12060605.

[32] Saranya, S., & Al-Mdallal, Q. M. (2021). Computational study on nanoparticle shape effects of Al_2O_3-silicon oil nanofluid flow over a radially stretching rotating disk. *Case Studies in Thermal Engineering*, 25, 100943. https://doi.org/10.1016/j.csite.2021.100943.

[33] Divya, S., Alessa, N., Eswaramoorthi, S., & Loganathan, K. (2022). Thermally radiative Darcy-Forchheimer flow of Cu/Ag nanoliquid in water past a heated stretchy sheet with magnetic and viscous dissipation impacts. *Symmetry*, 15(1), 16. https://doi.org/10.3390/sym15010016.

[34] Rehman, K. U., Malik, M. Y., Zahri, M., & Tahir, M. (2018). Numerical analysis of MHD Casson Navier's slip nanofluid flow yield by rigid rotating disk. *Results in Physics*, 8, 744–751. https://doi.org/10.1016/j.rinp.2018.01.017.

[35] Mustafa, M. (2017). MHD nanofluid flow over a rotating disk with partial slip effects: Buongiorno model. *International Journal of Heat and Mass Transfer*, 108, 1910–1916. https://doi.org/10.1016/j.ijheatmasstransfer.2017.01.064.

Chapter 12

Modeling of hybrid Sisko nanofluid over a radially stretching surface with suction

Sanjalee Maheshwari
NIT Hamirpur

Ankita Bisht
Amity University Punjab

Arvind Singh Bisht
NIT Hamirpur

12.1 INTRODUCTION

In recent decades, the research community has shown significant interest in fluid flow around a solid object, particularly when the fluid comes to a complete stop, as it finds numerous applications in various fields such as aerospace, automobile, and chemical engineering. This particular type of flow is referred to as stagnation-point flow. The exploration of stagnation-point flow past an infinite surface was initiated by Hiemenz [1], which later attracted many researchers to investigate this phenomenon under various hydrodynamic conditions. Howarth [2] improved the findings of Hiemenz [1], while Eckert [3] extended the work of Hiemenz [1] by considering the thermal energy in the stagnation-point flow. Kapur and Srivastava [4] proposed a two-dimensional mathematical model for the stagnation-point flow of a power-law fluid past a rigid surface. Building upon this work, Mahapatra et al. [5] extended the model proposed by Kapur and Srivastava [4] for the magnetohydrodynamic stagnation-point flow of a power-law fluid over the expanding surface. Subsequently, numerous studies have been reported in the literature, investigating stagnation-point flow over an expanding surface under different hydrodynamic perspectives. Some pioneering works in this field are found in references [6–10].

Several of the above-mentioned works focus on the stagnation-point flow of different types of non-Newtonian fluids, including power-law and grade fluids. Among these non-Newtonian fluid models, the Sisko fluid model proposed by Sisko [11] is widely used and recognized as a standalone model in non-Newtonian fluid studies. The Sisko fluid model provides a description

 DOI: 10.1201/9781032712079-12

of the behavior of both non-Newtonian and Newtonian fluids and can be seen as the generalization of the power-law fluid model. Sisko fluids are commonly encountered in nature and found diverse applications in areas such as drug administration, metal casting, tissue adhesion, and coating. Khan and Shahzad [12] conducted numerical investigations on the stagnation-point flow of Sisko fluid over a stretching surface and found that Sisko fluid exhibits a much stronger velocity profile compared with Newtonian and power-law fluids when the ratio of free stream velocity to linear stretching sheet velocity is less than unity. Later, Khan et al. [13] studied the impact of nonlinear thermal radiation on the stagnation-point flow of Sisko fluid over a stretching cylinder. In another study, Khan et al. [14] quantified the amount of entropy generation resulting from the flow of a chemically reactive Sisko fluid containing suspended nanoparticles over a stretching surface at a stagnation point.

MHD flow, also known as magnetohydrodynamic flow, is a specialized branch of fluid dynamics that deals with the interaction between a conducting fluid and magnetic fields. The study of MHD flow has gained the attention of researchers in the past few decades due is potential applications in engineering such as magnetohydrodynamic power generation (generating electricity directly from hot, ionized gases using magnetic fields) and magnetohydrodynamic propulsion (using magnetic fields to accelerate a conducting fluid for propulsion in space). As of late, Pal and Mandal [15] numerically discussed the influence of a magnetic field on the stagnation-point flow of Siskonanofluid past a stretching sheet. Numerous other pioneering works in this area are well documented in reference [16–20].

In light of recent advancements in nanotechnology, a novel concept of hybrid nanofluids has emerged. These innovative fluids exhibit enhanced rheological, thermophysical, optical, and morphological characteristics. Hybrid nanofluids are prepared in the laboratory by suspending more than one type of nanoparticles in the same base fluid, leading to superior properties in terms of rheology, thermophysics, optics, and morphology. Extensive research has demonstrated that the use of hybrid nanofluids significantly improves the heat transmission rate across various applications. Compared with mono nanofluids, hybrid nanofluids are expected to replace regular nanofluids for several compelling reasons. They offer a broader absorption range, higher thermal conductivity, lower pressure drop, and reduced frictional losses and pumping power. Consequently, hybrid nanofluids have the potential to find applications in different sectors, including electronics cooling, automotive heat exchangers, power production, and solar collectors, among others, where they have been subject to intensive investigation. Adopting hybrid nanofluids over conventional heat transfer fluids can bring several advantages, such as improved stability, enhanced thermal conductivity, and increased heat transfer efficiency [21,22].

Hussain et al. [23] presented a study to inspect the behavior of convective boundary conditions on the heat transmission rate of hybrid nanofluids passing over an expanding rotating sheet. Madhesh and Kalaiselvam [24] performed a series of experiments to analyze the heat transfer characteristics of a hybrid nanofluid as a coolant. The impact of a magnetic field on the flow behavior of hybrid nanofluids over an expanding and contracting wedge was investigated by Waini et al. [25]. Waini et al. [26] explored the assisting/opposing flow of a hybrid nanofluid along a vertical surface. Khan et al. [27] presented a three-dimensional mathematical model to investigate the stagnation-point flow of a hybrid nanofluid past a circular cylinder.

Viscous dissipation, also known as viscous heating or viscous energy dissipation, is a phenomenon that occurs in fluid flows where mechanical energy is converted into a heat source due to internal friction within the fluid. When a fluid flows past a solid surface or experiences velocity gradients within itself, adjacent fluid layers move at different speeds, resulting in shear stresses between these layers. This shear stress leads to energy losses within the fluid, causing a rise in temperature and an increase in internal energy. Thus, viscous dissipation can affect the temperature distribution and thermal characteristics of the fluid flow, and in some cases, it may lead to changes in boundary layer thickness and alter the flow patterns. Being motivated by the importance of viscous dissipation Zainal et al. [28] considered the influence of viscous dissipation and a transversal magnetic field while studying hybrid nanofluid flow over an expanding or contracting surface. Their findings revealed a significant increase in heat transmission rate and skin friction with the accumulation of nanoparticles. Using the single-phase nanofluid model.

The comprehensive literature review revealed that no previous research has investigated the stagnation-point flow of a Sisko hybrid nanofluid over a radially stretching surface, considering the effects of a magnetic field and viscous dissipation. To fill this research gap, a mathematical model is developed to study the flow of the hybrid Siskonanofluid over the radially stretching surface, taking into account the influence of a transversal magnetic field. The governing flow dynamics of the hybrid Siskonanofluid system are solved using the MATLAB bvp4c routine. The outcomes of the study are graphically presented and thoroughly discussed. The primary aim of this chapter is to analyze the thermal performance of the hybrid Sisko nanofluidic system under the specified conditions.

12.2 MATHEMATICAL MODELING

Consider 2D steady stagnation-point flow of Sisko hybrid nanofluid over a permeable radially stretching disk having nonlinear velocity $U_w = c^* r^{s^*}$ and the velocity of the free stream flow is $U_w = d^* r^{s^*}$ where c^* and d^* are constants and s^* is the stretching parameter. Figure 12.1 show a physical illustration of the problem, with the associated cylindrical polar coordinates

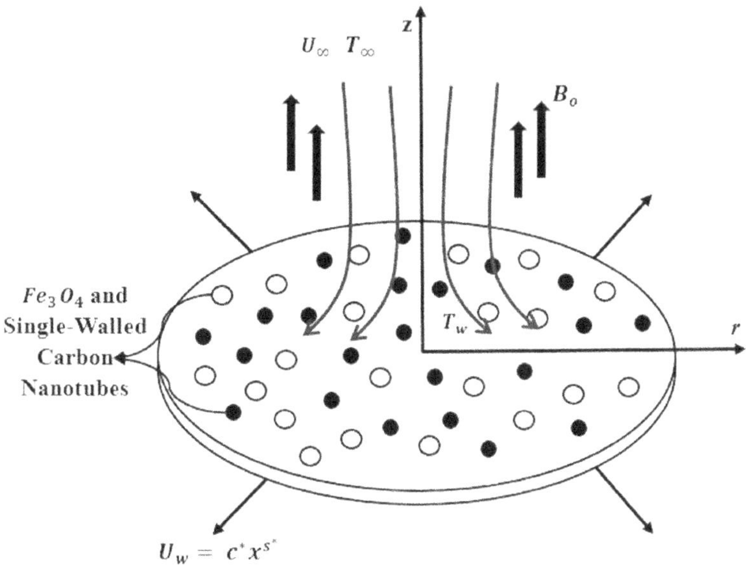

Figure 12.1 Physical sketch of the problem.

z and r. The coordinate z-axis is along normal to the disk, while the radial coordinate r-axis is along the disk's surface. The variable magnetic field with intensity B_0 is applied perpendicular to the surface. The sheet is at the constant temperature T_w; moreover, T_∞ represents the free stream fluid temperature when $z \to \infty$, such that $T_w > T_\infty$. The governing conservation equations of the model under study in terms of cylindrical polar coordinates (r, z), under the aforementioned presumptions [12,15,29,30] and using the hybrid nanofluid model given by Talebi and Salehi [31], are as follows:

$$\frac{\partial(r\tilde{u})}{\partial r} + \frac{\partial(r\tilde{w})}{\partial z} = C \tag{12.1}$$

$$\tilde{u}\frac{\partial\tilde{u}}{\partial r} + \tilde{w}\frac{\partial\tilde{u}}{\partial z} = U_\infty \frac{dU_\infty}{dx} + \frac{\mu_{hnf}}{\rho_{hnf}}\frac{\partial^2\tilde{u}}{\partial z^2} + \frac{b^*}{r\rho_{hnf}}\left(-\frac{\partial\tilde{u}}{\partial z}\right)^{n^*} + \frac{\sigma_{hnf}B_o^2}{\rho_{hnf}}(U_\infty - \tilde{u}) \tag{12.2}$$

$$\tilde{u}\frac{\partial T}{\partial r} + \tilde{w}\frac{\partial T}{\partial z} = \frac{\kappa_{hnf}}{(\rho Cp)_{hnf}}\frac{\partial^2 T}{\partial z^2} + \frac{1}{(\rho Cp)_{hnf}}\left[\mu_{hnf}\left(-\frac{\partial\tilde{u}}{\partial z}\right)^2 + b^*\left(-\frac{\partial\tilde{u}}{\partial z}\right)^{n^*+1}\right]$$

$$+\frac{\sigma_{hnf}B_o^2}{\rho_{hnf}}(U_\infty - \tilde{u})^2 \tag{12.3}$$

The physical boundary conditions are

$$\text{At} \quad z=0 \; : \tilde{u}=U_w=c^* x^{s^*}, \quad \tilde{w}=v_w, \quad T=T_w$$

and as

$$z \to \infty : \tilde{u} \to U_\infty = d^* x^{s^*}, \quad \tilde{w} \to 0, \quad T \to T_\infty, \tag{12.4}$$

where $\tilde{u}$ and $\tilde{w}$ denotes the nanofluid velocity in axial (x) and radial (r) directions, respectively.

In accordance with Khan et al. [30], the following dimensionless variables are used:

$$\psi(r,z) = U_w \, r^2 \, Re_{b^*}^{-1/(n^*+1)} f(\eta), \quad \eta = \frac{z}{r} Re_{b^*}^{1/(n^*+1)}, \quad \theta = \frac{T-T_\infty}{T_w - T_\infty} \tag{12.5}$$

where the stream function ψ for the velocity is given by

$$\tilde{u} = \frac{1}{r}\frac{\partial \psi}{\partial z}, \quad \tilde{w} = -\frac{1}{r}\frac{\partial \psi}{\partial r} \tag{12.6}$$

Now, on using equation (12.5), in equation (12.6), we get

$$\tilde{w} = -U_w Re_{b^*}^{\frac{-1}{(n^*+1)}} \left[\left(\frac{s^*(2n^*-1)+n^*+2}{n^*+1} \right) f(\eta) + \left(\frac{s^*(2-n^*)-1}{n^*+1} \right) \eta f'(\eta) \right] \tag{12.7}$$

where $Re_{b^*} = \dfrac{\rho_f r^{n^*} U_w^{2-n^*}}{b^*}$ and $Re = \dfrac{U_w r}{v_f}$ denotes the local Reynolds number.

The leading governing nonlinear differential equations (12.2) and (12.3) with boundary conditions (12.4) are transformed by self-similarity transformations (12.5) into the dimensionless form of the following ODEs:

$$\frac{n^*(-f'')^{n^*-1}}{\rho_{hnf}/\rho_f} f''' + \frac{\mu_{hnf}/\mu_f}{\rho_{hnf}/\rho_f} A^* f''' + \left(\frac{s^*(2n^*-1)+n^*+2}{n^*+1} \right) ff'' + \chi^2 - s^* f'^2$$
$$+ \frac{\sigma_{hnf}/\sigma_f}{\rho_{hnf}/\rho_f} M^*[\chi - f'] = 0 \tag{12.8}$$

$$\theta'' + Pr^* \left(\frac{s^*(2n^*-1)+n^*+2}{n^*+1} \right) \frac{(\rho Cp)_{hnf}/(\rho Cp)_f}{\kappa_{hnf}/\kappa_f}$$
$$f\theta' + \frac{\kappa_f}{\kappa_{hnf}} Ec^* Pr^* \left[A^*(-f'')^2 + (-f'')^{n^*+1} \right] + Ec^* Pr^* M^* [\chi - f']^2 = 0 \tag{12.9}$$

subjected to boundary conditions:

$$\begin{aligned} &\text{At} \quad \eta \to 0 : f(0) = f_w^*, \quad f'(0) = 1, \quad \theta(0) = 1, \\ &\text{as} \quad \eta \to \infty : f'(\eta) \to \chi, \quad \theta(\eta) \to 0, \end{aligned} \tag{12.10}$$

The physical parameters involved in equations (12.8)–(12.10) are defined as the ratio of rates, that is, the ratio of free stream velocity to the velocity of linear stretching disk (χ), Sisko material parameter (A^*), Prandtl number (Pr^*), magnetic parameter (M^*), Eckert number (Ec^*), suction parameter (f_w^*) where

$$\chi = \frac{d^*}{c^*}, \quad A^* = \frac{Re_{b^*}^{2/(n^*+1)}}{Re_{a^*}}, \quad Pr^* = \frac{r\,U_w}{\alpha_f} Re_{b^*}^{-2/(n^*+1)}, \quad M^* = \frac{\sigma_f B_0^2\, x}{\rho_f\, U_w},$$

$$Ec^* = \frac{U_w^2}{Cp_f\,(T_w - T_\infty)}, \quad f_w^* = \frac{-v_w Re_{b^*}^{1/(n^*+1)}}{\left(\dfrac{s^*(2n^*-1) + n^* + 2}{n^*+1}\right) U_\infty}$$

The physical quantities for the flow problem that is skin friction coefficient (C_f) and Nusselt number (Nu_x) (or heat transfer rate) are, respectively, defined as

$$C_f = \frac{2\,\tau}{\rho_f\, U_w^2}; \quad \tau = \left[\mu_{bnf} \frac{\partial \tilde{u}}{\partial z} - b^* \left(-\frac{\partial \tilde{u}}{\partial z}\right)^{n^*}\right]_{z=0}, \quad Nu = \frac{\kappa_{bnf}\, r}{\kappa_f (T_w - T_\infty)} \left.\frac{\partial T}{\partial z}\right|_{z=0}$$

After applying the similarity transformation to the above expression, we get

$$\frac{1}{2} C_f Re_{b^*}^{1/(n^*+1)} = A^* \frac{\mu_{bnf}}{\mu_f} F_\eta(0) - \left(-F_\eta(0)\right)^{n^*}$$

$$Re_{b^*}^{-1/(n^*+1)} Nu = -\frac{\kappa_{bnf}}{\kappa_f} \theta_\eta(0)$$

12.3 SOLUTION METHODOLOGY

The non-dimensional system of equations (12.8) and (12.9) together with boundary conditions (12.10) is highly nonlinear in nature and cannot be solved analytically. Therefore, the flow problem is solved numerically using the finite difference approach-based bvp4c routine of MATLAB. In this technique, nonlinear differential equations are transformed into unit-order differential equations. Numerical computations are carried out for the integration range ($\eta_{max} = 6$) that is found to be sufficient for the velocity

Table 12.1 Thermophysical laws for hybrid nanofluid [32]

Properties	*Hybrid nanofluid model*
Viscosity	$\mu_{hnf} = \dfrac{\mu_f}{(1-\phi_1-\phi_2)^{2.5}}$
Density	$\rho_{hnf} = \rho_f(1-\phi_1-\phi_2)+\rho_1\phi_1+\rho_2\phi_2$
Heat capacity	$(\rho Cp)_{hnf} = (\rho Cp)_f(1-\phi_1-\phi_2)+(\rho Cp)_1\phi_1+(\rho Cp)_2\phi_2$
Thermal conductivity	$\dfrac{\kappa_{hnf}}{\kappa_{bf}} = \dfrac{\kappa_2+2\kappa_f-2\phi_2(\kappa_{bf}-\kappa_2)}{\kappa_2+2\kappa_f+\phi_2(\kappa_{bf}-\kappa_2)}$ where $\dfrac{\kappa_{bf}}{\kappa_f} = \dfrac{\kappa_1+2\kappa_f-2\phi_1(\kappa_f-\kappa_1)}{\kappa_1+2\kappa_f+\phi_1(\kappa_f-\kappa_1)}$
Electrical conductivity	$\dfrac{\sigma_{hnf}}{\sigma_{bf}} = \dfrac{\sigma_2+2\sigma_f-2\phi_2(\sigma_{bf}-\sigma_2)}{\sigma_2+2\sigma_f+\phi_2(\sigma_{bf}-\sigma_2)}$ where $\dfrac{\sigma_{bf}}{\sigma_f} = \dfrac{\sigma_1+2\sigma_f-2\sigma_1(\sigma_f-\sigma_1)}{\sigma_1+2\sigma_f+\sigma_1(\sigma_f-\sigma_1)}$

Table 12.2 Thermophysical quantities of base fluid and nanoparticles at 300 K [33]

	Viscosity (μpas)	*Density* (ρ) *kg-m*$^{-3}$	*Specific heat* (Cp)*J/kg K*	*Thermal conductivity* (κ)*W/mK*	*Electrical conductivity* (σ) *S/m*
Blood	3.5×10^{-3}	1050	3900	0.492	0.8
SWCNTs	–	2600	425	6600	10^6
Iron oxide	–	5180	670	9.7	$10^{-5.99}$

and temperature profiles to converge asymptotically under the predefined boundary conditions. The iterations are carried out until the solutions for the desired level of accuracy, that is up to 10^{-6} is achieved to meet the convergence criterion. The thermophysical characteristics of the working fluid are calculated using the thermophysical laws given in Table 12.1. The numerical values of the thermophysical quantities of base fluid and nanoparticles are given in Table 12.2.

12.4 RESULTS AND DISCUSSION

In this section, the influence of significant flow governing parameters including the suction parameter, velocity of linear stretching sheet, Sisko material parameter, and magnetic field parameter on velocity distribution, temperature distribution, and quantities of engineering interest such as Nusselt number and skin friction coefficient has been briefly discussed graphically.

Figures 12.2–12.11 are delineated to study the variation in velocity and temperature profile with pertinent flow governing parameters while variation in Nusselt number and skin friction coefficient is shown in Figures 12.12 and 12.13. Changes in the velocity profile with the Sisko material parameter are presented in Figure 12.2. Since fluid viscosity and Sisko material parameter are inversely related to each other therefore the increase in Sisko

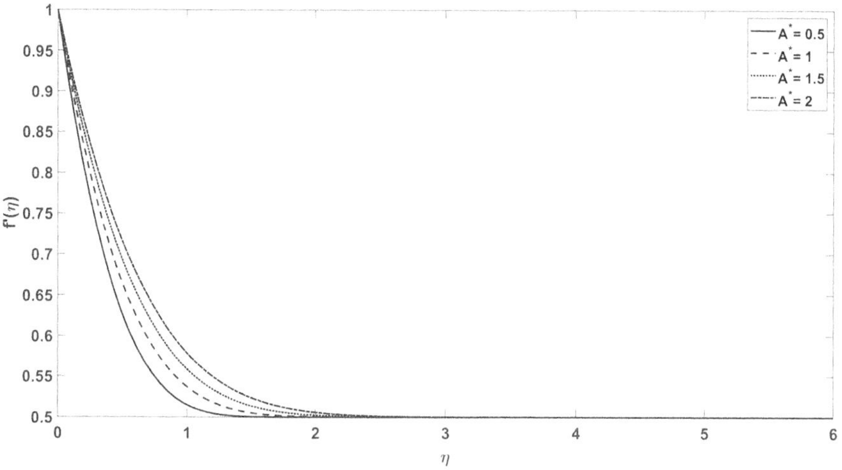

Figure 12.2 Variation of velocity profile with Sisko material parameter.

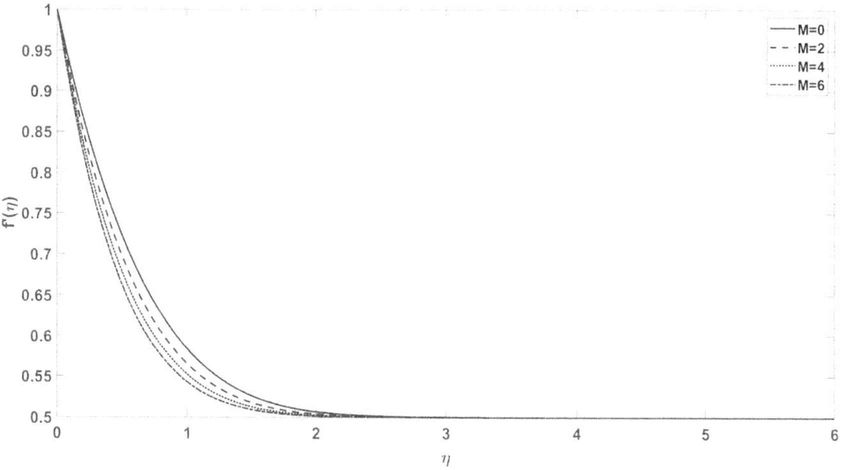

Figure 12.3 Variation of velocity profile with Magnetic parameter.

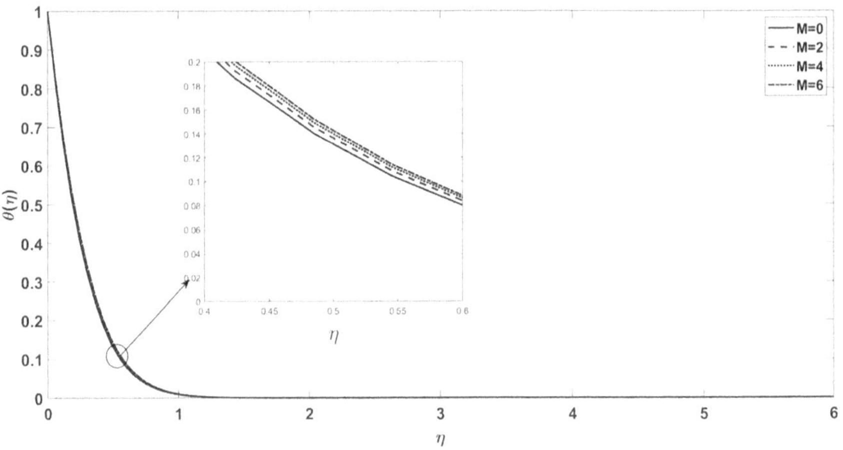

Figure 12.4 Variation of temperature profile with Magnetic parameter.

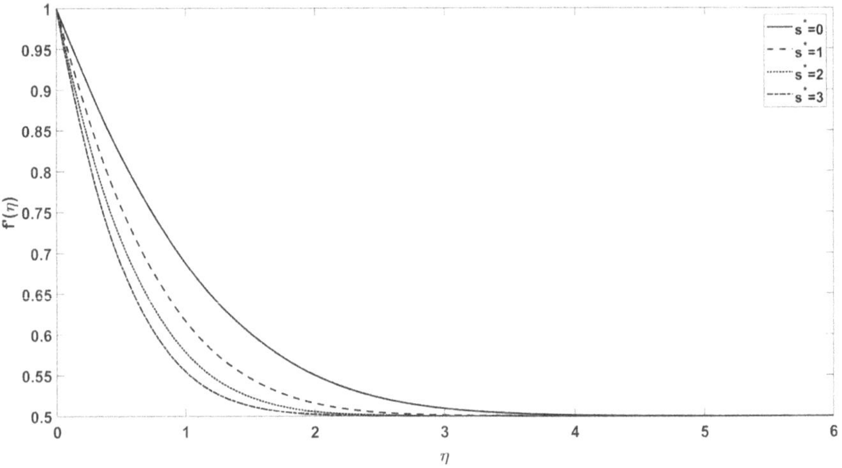

Figure 12.5 Variation of velocity profile with stretching parameter.

material parameter, decrease the fluid viscosity and consequently increases the fluid velocity near the boundary layer regime as shown in Figure 12.2.

The variation of magnetic field parameters on the velocity and temperature distribution profiles is shown in Figures 12.3 and 12.4, respectively. A decrease in the velocity profile with magnetic field parameter is reported in Figure 12.3. This outcome is physically anticipated because the increase in the magnetic field parameter intensifies the Lorentz force that acts in the reverse direction of flow. This Lorentz force exerts the resistance on the flow that leads to heat generation and consequently rise in the thermal distribution profile (Figure 12.4).

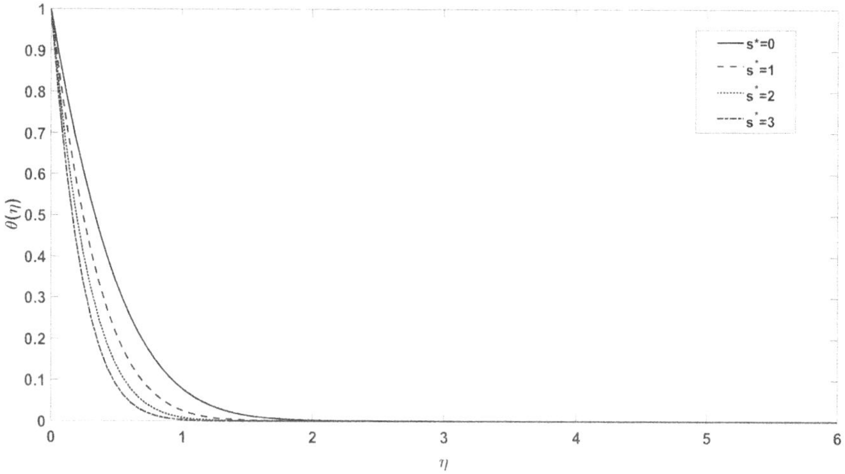

Figure 12.6 Variation of temperature profile with stretching parameter.

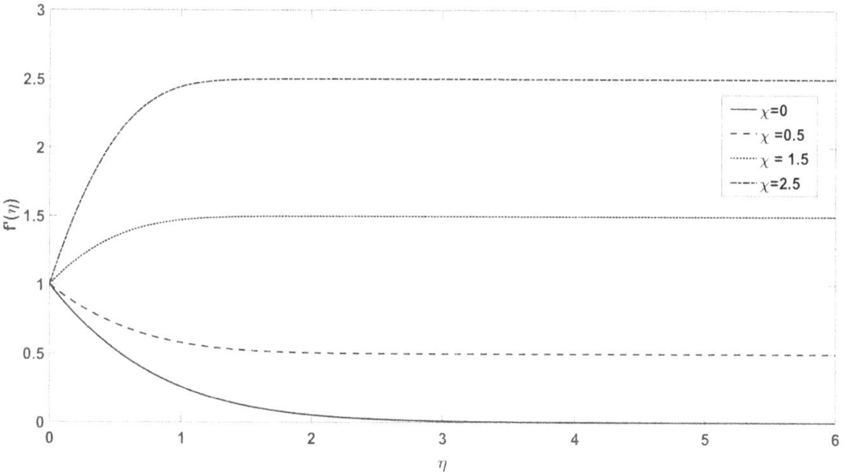

Figure 12.7 Variation of velocity profile with ratio of rates.

Figures 12.5 and 12.6 illustrate the variation in velocity and temperature profiles, respectively, as the stretching parameter is altered. From Figure 12.5 it is observed that with the increase in stretching parameter, radial velocity decreases and consequently momentum boundary layer thickness decreases. The same trend is observed for the temperature profile in Figure 12.6. It is reported that both velocity and temperature profiles asymptotically approach zero beyond a finite distance from the disk and decrease up to a finite distance from the sheet. Because of this characteristic, flow past stretching sheet finds application in the cooling of electronic equipments.

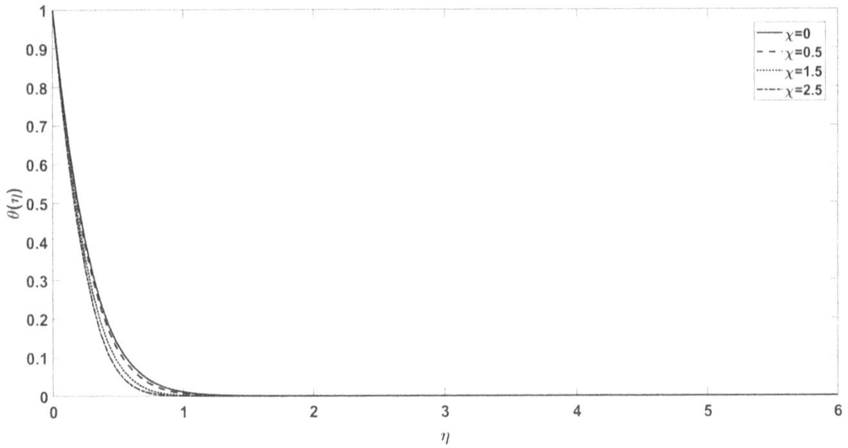

Figure 12.8 Variation of temperature profile with ratio of rates.

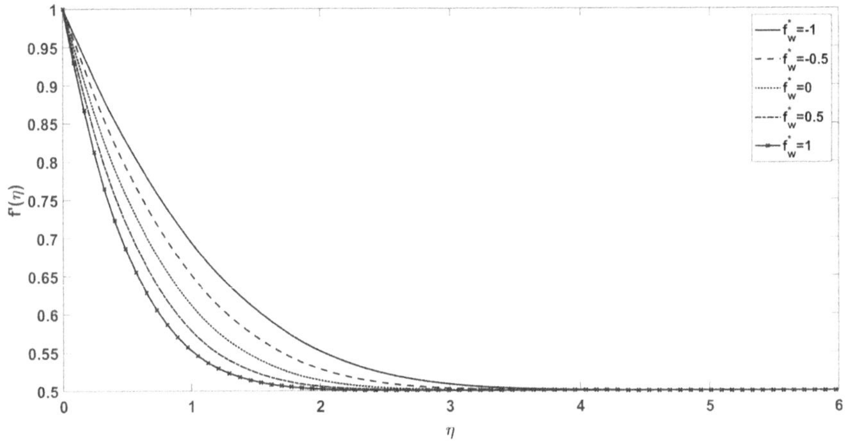

Figure 12.9 Variation of velocity profile with suction parameter.

The effect of the ratio of rates, that is, the ratio of free stream velocity to stretching sheet velocity on flow velocity and temperature profile is shown in Figures 12.7 and 12.8. Figure 12.7 shows that for a larger ratio of rates flow velocity increases in the boundary layer regime while momentum boundary layer thickness increases when the velocity of the linear stretching sheet dominates the free stream velocity. On the contrary, a decrease in the momentum boundary layer thickness is reported when free stream velocity is greater than the linear stretching sheet velocity. Further absence of momentum boundary layer is observed when the ratio rate is unit, that is, free stream velocity and linear stretching sheet velocity are equal.

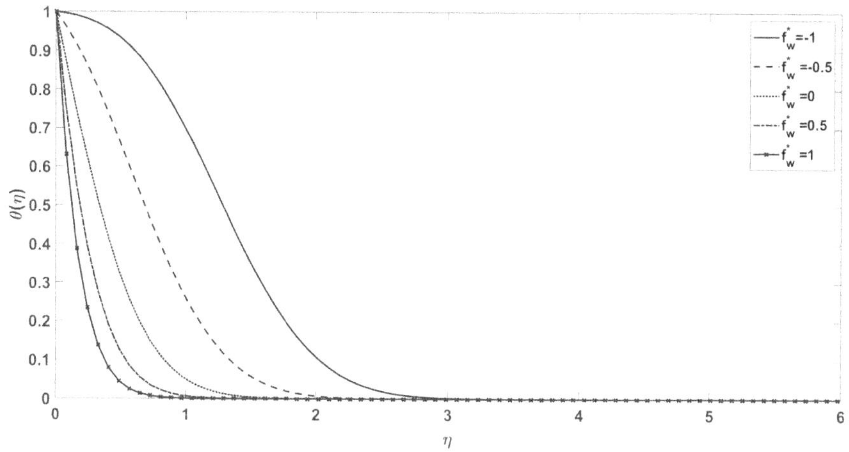

Figure 12.10 Variation of temperature profile with suction parameter.

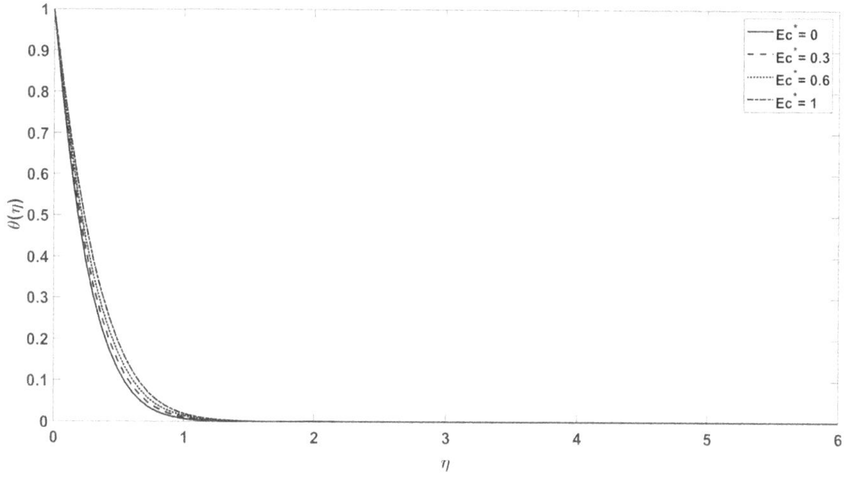

Figure 12.11 Variation of temperature profile with Eckert number.

A continuous decrease in the thermal boundary layer thickness with ratio rates is shown in Figure 12.8.

Figures 12.9 and 12.10 are plotted to observe the effect of suction parameters on velocity and temperature profile. Positive values of f_w^* corresponds to the phenomena of suction of fluid at the stretching surface. It is concluded from the plots that the velocity and temperature of the fluid decrease with the suction parameter near the boundary regime. These results are physically anticipated since suction pushes the warm fluid toward the boundary wall, which leads to an increase in the hybrid nanofluid viscosity and a decrease in wall shear stress.

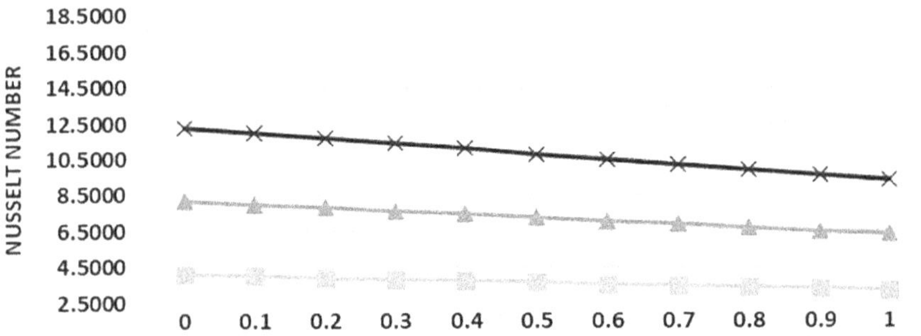

Ec*	0	0.1	0.2	0.3	0.4	0.5	0.6	0.7	0.8	0.9	1
M=2	4.087	3.955	3.822	3.690	3.557	3.425	3.292	3.160	3.028	2.895	2.763
M=1	4.094	3.982	3.870	3.758	3.646	3.534	3.423	3.311	3.199	3.087	2.975
M=0.5	4.097	3.996	3.895	3.795	3.694	3.593	3.492	3.391	3.290	3.190	3.089
M=0	4.101	4.011	3.922	3.833	3.743	3.654	3.565	3.475	3.386	3.297	3.207

EC*

Figure 12.12 Variation of Nusselt number with Eckert number and magnetic field parameter.

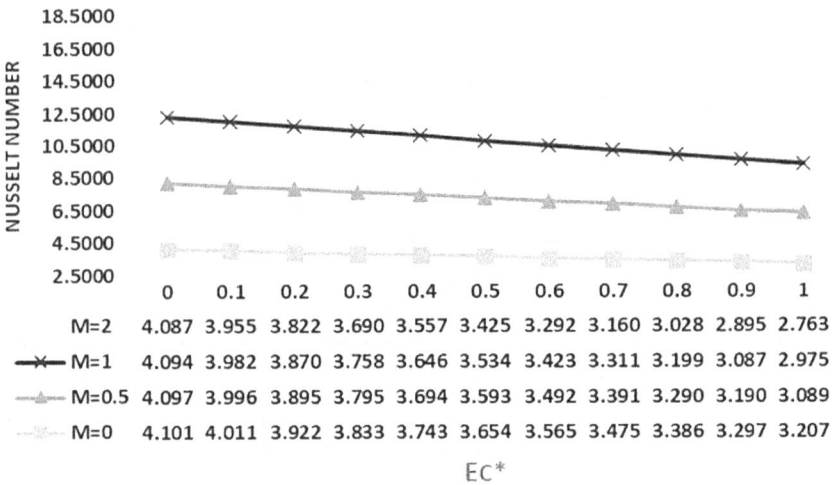

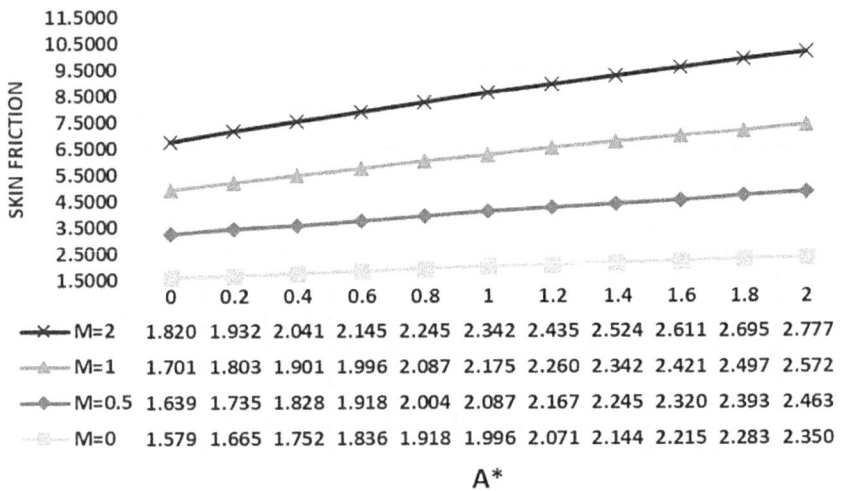

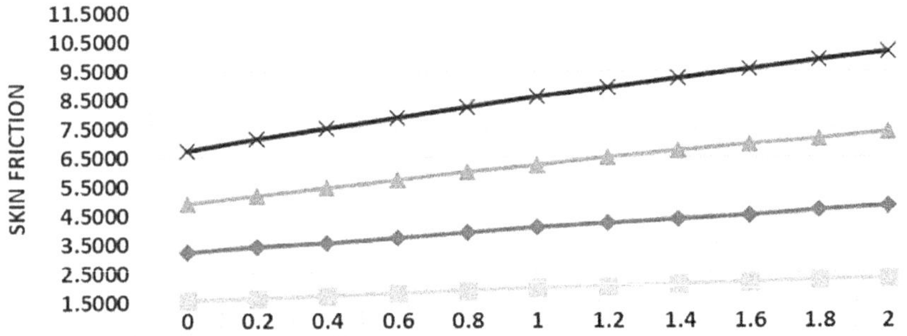

A*	0	0.2	0.4	0.6	0.8	1	1.2	1.4	1.6	1.8	2
M=2	1.820	1.932	2.041	2.145	2.245	2.342	2.435	2.524	2.611	2.695	2.777
M=1	1.701	1.803	1.901	1.996	2.087	2.175	2.260	2.342	2.421	2.497	2.572
M=0.5	1.639	1.735	1.828	1.918	2.004	2.087	2.167	2.245	2.320	2.393	2.463
M=0	1.579	1.665	1.752	1.836	1.918	1.996	2.071	2.144	2.215	2.283	2.350

A*

Figure 12.13 Variation of Skin friction coefficient with Sisko material parameter and magnetic field parameter.

The behavior of Eckert number on thermal distribution is presented in Figure 12.11. Physically Eckert number refers to the amount of dissipated heat through the internal friction during the mechanical to thermal energy conversion process. This thermal friction enhances the fluid temperature, and this decrease is graphically shown in Figure 12.11. Further effect of magnetic field parameter and Eckert number on Nusselt number is presented through the line graphs in Figure 12.12. With the increase in Eckert number, heat transfer rate decreases while with the increase in the magnetic field parameter, heat transfer rate decreases. This result is physically valid since with the increase in thermal dissipation heat transfer rate decreases and due to the resistance produced by Lorentz force heat transfer rate increases.

Figure 12.13 illustrates changes in the skin friction coefficient as the values of both the Sisko material parameter and the magnetic field parameter are varied. The graph clearly shows that an increase in the Sisko material parameter leads to an increase in the skin friction coefficient. This relationship can be explained by considering the definition of the Sisko material parameter, which represents the ratio of shear viscosity to the consistency index. As A^* increases, the viscous force decreases, resulting in a reduction of the thickness of the momentum boundary layer and an enhancement of the skin friction coefficient. In addition, it is observed that the skin friction coefficient also increases with the magnetic field parameter.

12.5 CONCLUDING REMARKS

The flow behavior and heat transfer characteristic of stagnation-point flow of hybrid Siskonanofluid over the radially stretching disk under the effect of viscous dissipation and magnetic field is numerically investigated, and outcomes are briefly discussed through plots in the current study. The key findings of the study are abridged as follows:

- A magnetic field helps to monitor the velocity and temperature profile. Strengthened magnetic field intensity slowdowns the flow while increasing the fluid temperature.
- Increase in the momentum boundary layer thickness reported with augmentation in Sisko material parameter. Further, with the increase in Sisko material parameter, the drag force exerted by the fluid on the stretching surface strengthened.
- Viscous dissipation transfers the thermal energy in the system and increases the fluid temperature, but this heat dissipates from the system reducing the heat transfer rate.
- Augmentation in the suction parameter results in an increase in the fluid viscosity near the stretching surface which results in the decrease in the velocity profile.
- The absence of a momentum boundary layer is observed when free stream velocity and linear stretching sheet velocity are equal.

NOMENCLATURE

A^*	Material parameter for Sisko fluid
b^*	Material constant for Sisko fluid
c^*	Constant

C_f	Local skin friction coefficient
f	Dimensionless stream function
M^*	Magnetic parameter
n^*	Power-law index
Nu	Local Nusselt number
Pr^*	Prandtl number
s^*	Stretching parameter
T	Temperature within the boundary layer (K)
$U_w(x)$	Stretching sheet velocity (m/s)
$\tilde{u}$	Axial velocity component (m/s)
$\tilde{v}$	Radial velocity component (m/s)
(r, z)	Space coordinates (m)
α	Thermal diffusivity (m^2/s)
η_∞	Edge of the boundary
ψ	Dimensional stream function
ν	Kinematic viscosity of nanofluid (m^2/s)
κ	Thermal conductivity (W/m-K)
ρ	Density of fluid (kg/m^3)
σ	Electrical conductivity (S/m)
θ	Dimensionless temperature

SUBSCRIPTS

hnf	Hybrid nanofluid
f	Base fluid
np	Nanoparticles
1	Single walled carbon nanotubes
2	Iron oxide

REFERENCES

[1] Hiemenz, K. (1911). Die Grenzschicht an einem in den gleichformigen Flussigkeitsstrom eingetauchten geraden Kreiszylinder. *Dinglers Polytechnisches Journal*, *326*, 321–324.

[2] Howarth, L. (1935). "On the calculation of steady flow in the boundary layer near the surface of a cylinder in a stream." *Aeronautical Research Committee Reports and Memoranda no 1632*, 16–32.

[3] Eckert, E. R. G. (1942). Die Berechnung des Warmeuberganges in der laminarenGrenzschichtumstromterKorper. *VDI Forschungsheft*, *416*, 1–24.

[4] Kapur, J. N., & Srivastava, R. C. (1963). Similar solutions of the boundary layer equations for power law fluids. *Zeitschrift für Angewandte Mathematik und Physik*, *14*(4), 383–389.

[5] Mahapatra, T. R., Nandy, S. K., & Gupta, A. S. (2009). Magnetohydrodynamic stagnation-point flow of a power-law fluid towards a stretching surface. *International Journal of Non-Linear Mechanics*, *44*(2), 124–129.
[6] Nazar, R., Amin, N., Filip, D., & Pop, I. (2004). Stagnation point flow of a micropolar fluid towards a stretching sheet. *International Journal of Non-Linear Mechanics*, *39*(7), 1227–1235.
[7] Sadeghy, K., Hajibeygi, H., &Taghavi, S. M. (2006). Stagnation-point flow of upper-convected Maxwell fluids. *International Journal of Non-Linear Mechanics*, *41*(10), 1242–1247.
[8] Mahmoud, M. A., & Waheed, S. E. (2012). MHD stagnation point flow of a micropolar fluid towards a moving surface with radiation. *Meccanica*, *47*, 1119–1130.
[9] Hayat, T., Shehzad, S. A., Alsaedi, A., & Alhothuali, M. S. (2012). Mixed convection stagnation point flow of Casson fluid with convective boundary conditions. *Chinese Physics Letters*, *29*(11), 114704.
[10] Mustafa, M., Mushtaq, A., Hayat, T., & Alsaedi, A. (2015). Model to study the non-linear radiation heat transfer in the stagnation-point flow of power-law fluid. *International Journal of Numerical Methods for Heat & Fluid Flow*, *25*(5), 1107–1119.
[11] Sisko, A. W. (1958). The flow of lubricating greases. *Industrial & Engineering Chemistry*, *50*(12), 1789–1792.
[12] Khan, M., & Shahzad, A. (2013). On stagnation point flow of Sisko fluid over a stretching sheet. *Meccanica*, *48*, 2391–2400.
[13] Khan, M., Malik, R., & Hussain, M. (2016). Nonlinear radiative heat transfer to stagnation-point flow of Sisko fluid past a stretching cylinder. *AIP Advances*, *6*(5), 055315.
[14] Khan, M. I., Hayat, T., Alsaedi, A., Qayyum, S., & Tamoor, M. (2018). Entropy optimization and quartic autocatalysis in MHD chemically reactive stagnation point flow of Sisko nanomaterial. *International Journal of Heat and Mass Transfer*, *127*, 829–837.
[15] Pal, D., & Mandal, G. (2020). Magnetohydrodynamic stagnation-point flow of Siskonanofluid over a stretching sheet with suction. *Propulsion and Power Research*, *9*(4), 408–422.
[16] Bisht, A., & Sharma, R. (2021). Entropy generation analysis in magnetohydrodynamicSiskonanofluid flow with chemical reaction and convective boundary conditions. *Mathematical Methods in the Applied Sciences*, *44*(5), 3396–3417.
[17] Bisht, A., & Sharma, R. (2021). Numerical investigation of Siskonanofluid over a nonlinear stretching sheet with convective boundary conditions. *International Journal for Multiscale Computational Engineering*, *19*(1), 41–54.
[18] Bisht, A., & Bisht, A. S. (2022). Radiative heat transfer due to solar radiation in MHD Siskonanofluid flow. *Heat Transfer*, *51*(8), 7411–7434.
[19] Hussaini, A. A., Abdullahi, I., Tata, A. A., & Musa, A. (2022). Numerical approach to determine the simultaneous influence of thermal radiation and chemical reaction over MHD stagnation-point flow of Siskonanofluid. *Iconic Research and Engineering Journals*, *5*(9), 2456–8880.

[20] Rizwana, R., Hussain, A., & Nadeem, S. (2021). Mix convection non-boundary layer flow of unsteady MHD oblique stagnation point flow of nanofluid. *International Communications in Heat and Mass Transfer*, *124*, 105285.

[21] Sidik, N. A. C., Jamil, M. M., Japar, W. M. A. A., & Adamu, I. M. (2017). A review on preparation methods, stability and applications of hybrid nanofluids. *Renewable and Sustainable Energy Reviews*, *80*, 1112–1122.

[22] Sarkar, J., Ghosh, P., & Adil, A. (2015). A review on hybrid nanofluids: recent research, development and applications. *Renewable and Sustainable Energy Reviews*, *43*, 164–177.

[23] Hussain, A., Alshbool, M. H., Abdussattar, A., Rehman, A., Ahmad, H., Nofal, T. A., & Khan, M. R. (2021). A computational model for hybrid nanofluid flow on a rotating surface in the existence of convective condition. *Case Studies in Thermal Engineering*, *26*, 101089.

[24] Madhesh, D., & Kalaiselvam, S. (2014). Experimental analysis of hybrid nanofluid as a coolant. *Procedia Engineering*, *97*, 1667–1675.

[25] Waini, I., Ishak, A., & Pop, I. (2020). MHD flow and heat transfer of a hybrid nanofluid past a permeable stretching/shrinking wedge. *Applied Mathematics and Mechanics*, *41*(3), 507–520.

[26] Waini, I., Ishak, A., Groşan, T., & Pop, I. (2020). Mixed convection of a hybrid nanofluid flow along a vertical surface embedded in a porous medium. *International Communications in Heat and Mass Transfer*, *114*, 104565.

[27] Khan, M. I., Shah, F., Waqas, M., Hayat, T., & Alsaedi, A. (2019). The role of γAl_2O_3-H_2O and γAl_2O_3-$C_2H_6O_2$ nanomaterials in Darcy-Forchheimer stagnation point flow: an analysis using entropy optimization. *International Journal of Thermal Sciences*, *140*, 20–27.

[28] Zainal, N. A., Nazar, R., Naganthran, K., & Pop, I. (2021). Viscous dissipation and MHD hybrid nanofluid flow towards an exponentially stretching/shrinking surface. *Neural Computing and Applications*, *33*, 1–11.

[29] Khan, U., Zaib, A., Ishak, A., Al-Mubaddel, F. S., Bakar, S. A., Alotaibi, H., & Aljohani, H. M. (2021). Computational modeling of hybrid Siskonanofluid flow over a porous radially heated shrinking/stretching disc. *Coatings*, *11*(10), 1242.

[30] Khan, U., Zaib, A., Shah, Z., Baleanu, D., & Sherif, E. S. M. (2020). Impact of magnetic field on boundary-layer flow of Sisko liquid comprising nanomaterials migration through radially shrinking/stretching surface with zero mass flux. *Journal of Materials Research and Technology*, *9*(3), 3699–3709.

[31] Takabi, B., & Salehi, S. (2014). Augmentation of the heat transfer performance of a sinusoidal corrugated enclosure by employing hybrid nanofluid. *Advances in Mechanical Engineering*, *6*, 147059.

[32] Kumar, S., & Sharma, K. (2022). Entropy optimized radiative heat transfer of hybrid nanofluid over vertical moving rotating disk with partial slip. *Chinese Journal of Physics*, *77*, 861–873.

[33] Sanjalee, Sharma, Y. D., & Yadav, O. P. (2022). Entropy generation in magnetohydrodynamics flow of hybrid Casson nanofluid in porous channel: lie group analysis. *International Journal of Applied and Computational Mathematics*, *8*(5), 247.

Chapter 13

Modeling of ternary nanofluid flow through a diverging porous channel

J. K. Madhukesh
Amrita School of Engineering, Amrita Vishwa Vidyapeetham

K. Vinutha
Davangere University

G.K. Ramesh
K.L.E. Society's J.T. College

Manjunathayya Holliyavar
K.L.E. Society's S.S.M.S College

13.1 INTRODUCTION

The term "nanofluid" refers to a type of fluid that is made up of nanoscale particles dispersed throughout a base fluid. Generally, nanoscale particles have diameters less than 100 nm. Including nano-sized particles in the carrier fluid will result in the rise of thermal efficiency. The idea of nanofluid was initially given by Choi [1] in 1995. Adding nanoscale particles to the carrier liquids enhances their capacity to offer heat conductivity. Due to their higher thermal properties, nanofluids are used in broad sectors, like biomedical, automotive, transportation, electrical, and technology, for a variety of purposes, including the distribution of drugs, actual time chemical tracking of brain function, and the elimination of tumors. Research has shown that nanofluids can independently have thermal conductivity levels that are significantly higher than the base fluid. Since the invention of this novel concept, researchers have been increasingly interested in the uses of nanoliquid. Khan et al. [2] numerically examined how viscous dissipation and thermophoretic factors affected a radiative mixed convective flow. Shah et al. [3] considered the impact of suction as well as dual extending on the motion of different hybrid-based nanoliquid. This view presents several reports, some listed in [4–6].

DOI: 10.1201/9781032712079-13

A hybrid nanofluid (HNF) is a liquid that blends two different nanoparticles with a base liquid, leading to higher thermal efficiency than traditional nanofluids. HNFs have superior thermal conductivity qualities than ordinary nanofluids. Furthermore, ternary nanofluids (T-NFs) are made up of three types of nano-scaled particles combined with a base liquid, improving their thermal properties even further. These fluids have particular features that make them suited for a wide range of utilization. T-NF has the potential to have even more excellent thermal conductivity than HNFs. It has numerous scopes in various fields, including the engineering, chemical sciences, biomedical field, nuclear power plants, automotive fields, electronic transportation, heat exchangers to reduce heat loss, cancer therapies, drug delivery, transportation, and coolant in vehicles. The effects of heat sink/source (H-S/S), Newtonian heating (N-H), and axisymmetric Darcy–Forchheimer circulation and energy distribution of hybrid nanoliquid over a tiny needle are examined by Ramesh et al. [7]. The micro-structural slip performance of TNF (ternary nanoliquid) motion across an elastic surface is explored by Madhukesh et al. [8]. Ramesh et al. [9] examined how a T-NF behaves on a slippery surface while taking into account the H-S/S, suction, slip effect, and convective boundary condition. Shukla et al. [10] studied the impact of elasticity deformation on a water-based hybrid nanoliquid of carboxymethyl cellulose on a stretched sheet with an induced magnetic field. Animasaun et al. [11] examined the thermal characterization and evaluation of unsteady water-based T-NF on wedges in detail. The research looked precisely at the impact of low and high magnitudes of nanoparticle's thermal conductivity on bioconvection and surface-extending velocity.

When a conductive liquid moves in a magnetic field, electrical currents are generated that act as a mechanical force and control how the fluid moves. Flows of liquid metal with high magnetic field effects can be found in several fusion reactor components. The study of the electromagnetic structure and characteristics of electrically conductive fluids is done in the discipline of mechanics known as "magnetohydrodynamics" (MHDs) to understand how conducting fluids move in a strong magnetic field. Manufacturing, healthcare, finance, cooling systems, magnetic coils, accelerators, heat exchangers, technology, and mixing chambers are some sectors where the magnetic field effect can be seen. So many studies on the magnetic field effect have been conducted. Manohar et al. [12] examined the thermal behavior of carbon hybrid nanotubes in three distinct carrier liquids with a porous, cylindrical fin considering magnetic field. The consequences of MHD movement and Hall currents on boundary layer circulation and heat transfer across three-dimensional stretched surfaces were examined by Ferdows et al. [13]. Using the nanoliquid model, Swain et al. [14] examined the impacts of magnetic energy, comprising the Lorentz force, as well as the effect of an exponential space-dependent heat source on nano-based liquid. Khan et al. [15] used the Koo–Kleinstreuer–Li relationship and a water-based nanoliquid to scrutinize the radiative axisymmetric motion of Casson liquid in the context of a magnetic effect. Madhukesh et al. [16] explored the impact of magnetic dipoles on the circulation of an extended surface of Maxwell HNF. Das et al. [17] investigated the impact of the double layer of electricity (DLE) on simultaneous magneto-convection in a toward channel using a T-NF.

The electromagnetic radiation that a substance produces, because it generates heat, is known as thermal radiation (TR), and its characteristics depend on the material's temperature. TR research might have uses in a variety of sectors, including electrical energy, food production, solar energy sheets, weapon science and technology, nuclear power plants, turbines for gas, and aeronautical engineering. Due to their numerous uses, both heat transfer and flow systems within the range of nonlinear TR perform an important part in science and technology. Ramesh et al. [18] explored thermophoretic particle deposition and TR in the movement of a nanofluid composed of a hybrid material including carbonate nanotubes around a rotating sphere. Abbas et al. [19] investigated the effect of a magnetic field, radiation, and gravity on the distribution of temperatures within a solid sphere using numerical techniques. Zeeshan et al. [20] investigated the two-dimensional nanofluid movement affecting a porous stretched layer with nonlinear TR and slip behavior at the boundary enclosing energy viewpoint. Khan et al. [21] investigated the mixed convection of liquid-containing graphene oxide nanoparticles down a vertical plate while taking TR into account. Alzahrani et al. [22] studied the effect of TR on the transmission of heat in a Casson nanofluid with plane wall jet motion, suction, and a slip boundary condition. The numerical solution of Maxwell–Sutterby nanofluid motion inside a stretched sheet with TR, exponential heat source/sink, and bioconvection was addressed by Zeeshan et al. [23].

A convergent symmetrical flow exists for each definite Reynolds value and any channel deviation, and two symmetrical irrotational boundary layer movements develop on the channel walls. On the other hand, for a given channel angle and Reynolds numbers below a threshold value, a fully divergent flow exists. If the Reynolds number is higher than the critical value, a backflow region forms in the divergent movement close to one of the channel walls. Boundary layer fluxes in convergent and divergent channels are essentially dissimilar from an existing perspective. The first scenario may frequently exist, while the other one is less common, and boundary layer separation can happen in the backflow zone based on the Reynolds number. Divergent channels were suggested and shown to be useful for increasing heat dissipation capacity and lowering pressure drop. Ramesh et al. [24] scrutinized how ternary HNFs behave in a stretched convergent/divergent channel when a porous material and a heat source/sink are present. Xu et al.

[25] studied the multiphase non-Newtonian flow over a diverging channel. Banerjee et al. [26] studied the boundary layer in a diverging porous channel with a non-Newtonian Casson liquid, including temperature transmission if suction/blowing and viscous dissipation are present. Nazeer et al. [27] studied the impact of a magnetic field and tiny metallic particles on the multiphase circulation of third-grade liquid in a diverging channel using an analytical solution. Verma et al. [28] studied the possible scenario of boundary layer nanoliquid movement along a divergent channel in a porous media with mass suction/injection using a numerical method. Isaev et al. [29] studied the computational and physical analysis of turbulent movement across a divergent channel containing a vortex cell.

With above-mentioned works, no one worked to investigate the two-dimensional, laminar, boundary layer flow of ternary HNFs in the existence of a magnetic field and TR across a divergent channel. The study further aims to analyze thermal transfer investigated on the ternary HNF flow of ternary-based nanofluid across a divergent channel by considering TR impact. The reduced ODEs and BCs are solved numerically with the help of computational software, and essential dimensionless constraints and engineering factors are discussed in detail. The outcomes of the work can be used in thermal management to improve efficiency, components of electronics, chemicals and thermal energy systems, and biomedical applications.

13.2 MATHEMATICAL FORMULATION

Consider a continuous, two-dimensional, laminar, boundary layer circulation of ternary HNFs in addition to magnetic field across a divergent channel. (u,v) pairs denotes the velocity components, and (x,y) denotes the Cartesian coordinates (see Figure 13.1). Further, TR is included in the temperature equation. The walls are related to injection/suction. The flow exists symmetrically in the central line of the channel. The angle between the channel is α. The variable magnetic field is introduced, and it is in the form $B = B_0 l / x$.

Governing equations related to the present problem may be written as follows (see [26])

$$\frac{\partial v}{\partial y} + \frac{\partial u}{\partial x} = 0, \tag{13.1}$$

$$\frac{\partial u}{\partial y} v + u \frac{\partial u}{\partial x} = -\frac{1}{\rho_{thnf}} \left(\frac{\partial P}{\partial x} \right) + \upsilon_{thnf} \frac{\partial^2 u}{\partial y^2} - \frac{\sigma_{thnf}}{\rho_{thnf}} B^2 u, \tag{13.2}$$

$$-\frac{1}{\rho_{thnf}} \left(\frac{\partial P}{\partial y} \right) = 0, \tag{13.3}$$

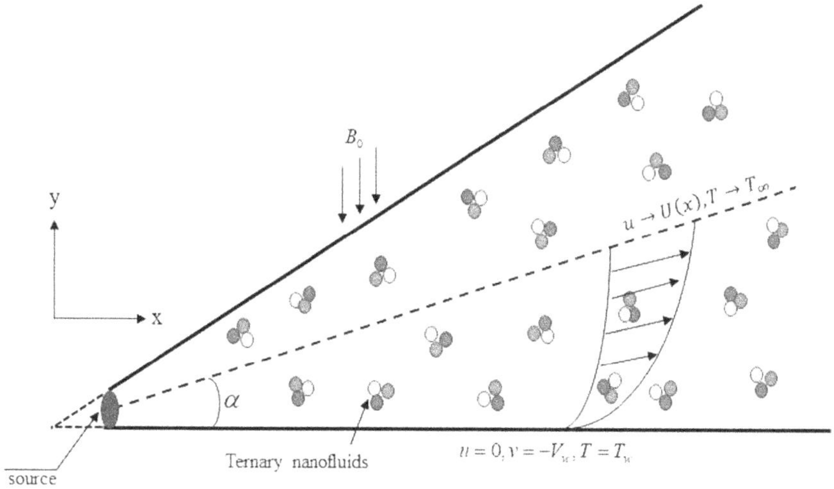

Figure 13.1 Geometrical representation of the problem.

$$u\frac{\partial T}{\partial x}+v\frac{\partial T}{\partial y}=\frac{k_{thnf}}{\left(\rho C_p\right)_{thnf}}\frac{\partial^2 T}{\partial y^2}-\frac{1}{\left(\rho C_p\right)_{thnf}}\frac{\partial q_r}{\partial y}. \tag{13.4}$$

The associated BCs are

$$\left.\begin{aligned} &u=0, \quad v=-V_w(x), \quad T=T_w : y=0,\\ &u \to U(x)[x>0] T \to T_\infty : y \to \infty. \end{aligned}\right] \tag{13.5}$$

From the above equations, T, υ_{thnf}, μ_{thnf}, ρ_{thnf}, σ_{thnf}, $\left(\rho C_p\right)_{thnf}$, and k_{thnf} correspond to temperature, kinematic fluid viscosity, viscosity of the fluid, density of the fluid, electrical conductivity, heat capacity, and thermal conductivity. U is the variable free steam velocity and is given below:

$$U(x)=\frac{Q}{\alpha x}=\frac{U^* l}{x}\left(U^*>0\right). \tag{13.6}$$

V_w is the mass suction/injection velocity.

$$V_w(x)=\frac{S_1}{x}\sqrt{\frac{Q\upsilon_f}{\alpha}}. \tag{13.7}$$

where U^* is the characteristic velocity, l indicates the characteristic length along the motion direction, Q is constant, α is the angle between channels, and S_1 is the mass suction/injection parameter (S_1 is >0 for suction and <0

for injection), P is the pressure gradient, and on determining the order of two momentum equation, it is given $|\partial P/\partial x| >> |\partial P/\partial y|$ and we concluded that P term is a dependent of x only and it is explained in equation (13.3). It is derived from equation (13.2) using the free stream region.

$$-\frac{1}{\rho_{thnf}}\frac{\partial P}{\partial x} = U\frac{dU}{dx}. \tag{13.8}$$

Eliminating $\partial P/\partial x$ from equations (13.2) and (13.8), we get

$$u\frac{\partial u}{\partial x} + v\frac{\partial u}{\partial y} = U\frac{dU}{dx} + \upsilon_{thnf}\frac{\partial^2 u}{\partial y^2} + \frac{\sigma_{thnf}B^2}{\rho_{thnf}}(U-u). \tag{13.9}$$

and q_r represents the nonlinear TR defined using Rosseland's approximation for radiation (see [19]).

$$q_r = \frac{-4\sigma *}{3k*}\frac{\partial T^4}{\partial y}. \tag{13.10}$$

Here T^4 is the linear function of temperature. And it is given by

$$T^4 \simeq 4T_\infty^3 T - 3T_\infty^4. \tag{13.11}$$

Substitute (13.11) in (13.10), we get

$$q_r = -\frac{16\sigma^*}{3k^*}T_\infty^3\frac{\partial T}{\partial y}. \tag{13.12}$$

By substituting equation (13.12) in (13.4), then the equation becomes

$$u\frac{\partial T}{\partial x} + v\frac{\partial T}{\partial y} = \left(\frac{k_{thnf}}{(\rho C_p)_{thnf}} + \frac{16\sigma^*}{3k*(\rho C_p)_{thnf}}T_\infty^3\right)\frac{\partial^2 T}{\partial y^2}. \tag{13.13}$$

where k^* is the absorption coefficient and σ^* is the Stefan–Boltzmann coefficient.

The following are the similarity variables for the present problem:

$$\left.\begin{array}{l} \eta = \frac{y}{x}\sqrt{\frac{Q}{\alpha\upsilon_f}}, \quad u = \frac{Q}{\alpha x}f(\eta), \\ v = \left(\frac{1}{x}\right)\sqrt{\frac{Q}{\alpha\upsilon_f}}\left[\eta f(\eta) - S\right], \quad \theta = \frac{T - T_\infty}{T_w - T_\infty}. \end{array}\right\} \tag{13.14}$$

Substituting equation (13.14), continuity equation (13.1) is identically satisfied, and equations (13.9) and (13.13) are reduced to

$$\frac{1}{\delta_1\delta_2}f'' - \left(\frac{\sigma_{thnf}}{\sigma_f\delta_2}\right)M^*(f-1) - (1-f^2) + fS_1 = 0, \tag{13.15}$$

$$\frac{1}{P_r}\theta''\left(\frac{k_{thnf}}{k_f} + \frac{4}{3}R_1\right) + \delta_3 S_1\theta' = 0. \tag{13.16}$$

where

$$\delta_1 = (1-\phi_3)^{2.5}(1-\phi_2)^{2.5}(1-\phi_1)^{2.5},$$

$$\delta_2 = \phi_3\frac{\rho_3}{\rho_f} + \left((1-\phi_2)\left(\phi_1\frac{\rho_1}{\rho_f} + (1-\phi_1)\right) + \phi_2\frac{\rho_2}{\rho_f}\right)(1-\phi_3),$$

$$\delta_3 = (1-\phi_3)\left((1-\phi_2)\left((1-\phi_1) + \phi_1\frac{(\rho C_p)_1}{(\rho C_p)_f}\right) + \phi_2\frac{(\rho C_p)_2}{(\rho C_p)_f}\right) + \phi_3\frac{(\rho C_p)_3}{(\rho C_p)_f}.$$

Reduced BCs are as follows:

$$\left.\begin{array}{ll} f(0)=0, & \theta(0)=1 \\ f(\infty)\to 1, & \theta(\infty)\to 0 \end{array}\right\} \tag{13.17}$$

where M^*, P_r, and R_1 are represented by the magnetic field parameter, Prandtl number, and radiation parameter, respectively.

$$M^* = \frac{\sigma_f B_0^2 \alpha l^2}{\rho_f Q}, \quad P_r = \frac{\mu_f (C_p)_f}{k_f}, \quad R_1 = \frac{4\sigma^* T_\infty^3}{k^* k_f}. \tag{13.18}$$

Thermophysical properties of ternary HNFs (see [17])

$$\mu_{thnf} = \delta_1\mu_f,$$

$$\frac{\rho_{thnf}}{\rho_f} = \delta_2,$$

$$\frac{(\rho C_p)_{thnf}}{\rho_f C_{pf}} = \delta_3,$$

$$\frac{k_{thnf}}{k_{hnf}}=\frac{k_3-\phi_3 2\left(k_{hnf}-k_3\right)+2k_{hnf}}{k_3+\phi_3\left(k_{hnf}-k_3\right)+2k_{hnf}},$$

$$\frac{k_{hnf}}{k_{nf}}=\frac{k_2-2\phi_2\left(k_{nf}-k_2\right)+2k_{nf}}{k_2+2k_{nf}+\phi_2\left(k_{nf}-k_2\right)},$$

$$\frac{k_{nf}}{k_f}=\frac{k_1-2\phi_1\left(k_f-k_1\right)+2k_f}{k_1+\phi_1\left(k_f-k_1\right)+2k_f}.$$

$$\frac{\sigma_{nf}}{\sigma_f}=\frac{\sigma_1-2\phi_1\left(\sigma_f-\sigma_1\right)+2\sigma_f}{\sigma_1+2\sigma_f+\phi_1\left(\sigma_f-\sigma_1\right)},$$

$$\frac{\sigma_{hnf}}{\sigma_{nf}}=\frac{\sigma_2-2\phi_2\left(\sigma_{nf}-\sigma_2\right)+\sigma_{nf}2}{\sigma_2+2\sigma_{nf}+\phi_2\left(\sigma_{nf}-\sigma_2\right)},$$

$$\frac{\sigma_{thnf}}{\sigma_{hnf}}=\frac{\sigma_3-2\phi_3\left(\sigma_{hnf}-\sigma_3\right)+2\sigma_{hnf}}{\sigma_3+2\sigma_{hnf}+\phi_3\left(\sigma_{hnf}-\sigma_3\right)}.$$

Here k-thermal conductivity, μ-dynamic viscosity, ρ-density, σ is the electrical conductivity and ϕ-solid fraction, C_p-heat capacity. In the above expression, $\phi_3=0$ it will reduce for HNF and $\phi_3=\phi_2=0$ reduces to nanofluid expression.

The essential engineering coefficients and their simplified form are given below (see [28])

$$Cf_x=\frac{\tau_w}{\rho_f U^2},\quad Nu_x=\frac{xq_w}{k_f\left(T_w-T_\infty\right)}. \tag{13.19}$$

where

$$\tau_w=\mu_{thnf}\left(\frac{\partial u}{\partial y}\right)_{y=0},\quad q_w=-k_{thnf}\left(\frac{\partial T}{\partial y}\right)_{y=0}. \tag{13.20}$$

Using equation (13.20), equation (13.19) reduces to the form,

$$Cf_x\left(Re\right)^{1/2}=\frac{1}{\delta_1}f'(0),\quad\text{and}\quad Nu_x\left(Re\right)^{-1/2}=-\frac{k_{thnf}}{k_f}\theta'(0). \tag{13.21}$$

Here $Re_x=\dfrac{Ux}{\upsilon_f}$ denotes the local Reynolds number.

13.3 NUMERICAL SECTION

The present section will focus on the methodology to obtain the solution of the reduced ODEs stated in the equations (13.15) and (13.16) and BCs in (13.17). Since the obtained equations are highly nonlinear and two point in nature. To obtain the solution of the resultant equations, we can reduce them into first-order system. For this, we take substitution as follows:

$$\begin{bmatrix} f, f' \\ \theta, \theta' \end{bmatrix} = \begin{bmatrix} P_1, P_2 \\ P_3, P_4 \end{bmatrix} \tag{13.22}$$

Using substitutions defined in (13.22), the equations (13.15) and (13.16) take the following form:

$$f'' = -\left(\left(\frac{\sigma_{thnf}}{\sigma_f \delta_2} \right) M^* (1 - \mathrm{P}_1) - \left(1 - \mathrm{P}_1^2\right) + S_1 \mathrm{P}_2 \right) \delta_1 \delta_2, \tag{13.23}$$

$$\theta'' = -P_r \delta_3 S_1 P_4 \Big/ \left(\frac{k_{thnf}}{k_f} + \frac{4}{3} R_1 \right). \tag{13.24}$$

$$\eta = 0 : P_1(0) = 0, \quad P_3(0) = 1, \quad \eta \to \infty : P_1(\infty) \to 1, \quad P_3(\infty) \to 0. \tag{13.25}$$

The resultant equations are numerically simplified by utilizing RKF-45 (Runge Kutta Fehlberg 4th 5th) method and the missing boundary conditions are guessed by implementing the shooting technique. The error tolerance is set about 10^{-6} during the computation and step size is taken 0.01. The thermophysical properties and thermophysical characteristics defined in the Table 13.1 are utilized. The stated numerical technique is compared with the work of [32] and found good agreement (see Table 13.2).

Table 13.1 Thermophysical properties of nanofluids and base fluid (see [30,31])

	ρ	C_p	k	σ	Pr
EG50% + H_2O	1056	3288	0.425	0.00509	29.86
$CoFe_2O_4$	4907	700	3.7	1.1×10^7	–
Fe_3O_4	5180	670	9.7	0.74×10^6	–
ZnO	5606	514	23.4	1.587	–

Table 13.2 Validation of the present work results with [32] for $f(\eta)$ in the absence of nanoparticles

η	*[32]*	*Present study results*
0.1	0.14930315	0.14930319
0.5	0.54338264	0.54338271
1.0	0.78295428	0.78295432
2.0	0.94854987	0.94854996
3.0	0.98749489	0.98749498

13.4 RESULTS AND DISCUSSION

This part focuses on presenting the study's innovative numerical findings. As described in the preceding section, the reduced ordinary differential equations (ODEs) are obtained with prescribed numerical approach. The study emphasizes the importance of important dimensionless factors such as S_1 (suction $(S_1 > 0)$ /injection $(S_1 < 0)$ parameter), M^* (magnetic parameter), and R_1 (radiation parameter), as well as their effect on the associated profiles. In addition, the engineering component is extensively examined. The findings are compared for both suction and injection instances, providing a thorough understanding of the influence of these factors on flow behavior. The results give important insights into the thermal properties and engineering consequences of T-NF flow through a diverging channel.

Figure 13.2 shows the disparity of the $f(\eta)$ for improved values of suction/injection constraint. As the values of S_1 increased, the velocity also improved in both cases. As S_1 values improve, it significantly enhances the velocity of the liquid within the divergent channel. However, it is worth noting that the velocity in the injection circumstance is smaller than in the suction scenario. This implies that the existence of injection reduces flow velocity to some amount, resulting in slightly reduced velocities than in the suction situation.

Figure 13.3 displays the fluctuation of $\theta(\eta)$ for numerous values of suction/injection constraints. As S_1 improves the thermal profile shows declination. This suggests that increased suction or injection levels cause a decrease in heat within the diverging channel. It is found that, in injection case it is seen that thermal distribution is comparatively less when compared with the suction scenario. This result shows that more cooling of the liquid is seen in injection case, this results in reduction of thermal profile than as seen suction condition.

Figure 13.4 illustrates the variation of $f(\eta)$ with increased M^* for suction and injection situations. The $f(\eta)$ shows improvement for both the cases with rise in M^*. This signifies that an improved magnetic field will uplift liquid velocity in the diverging channel. In the existence of M^*, velocity is larger in the suction case when compared with the injection case.

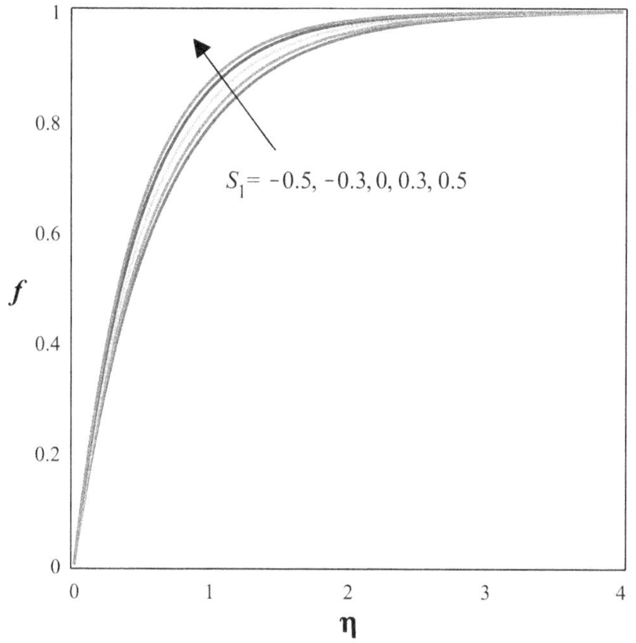

Figure 13.2 Varying $f(\eta)$ for several values of S_1.

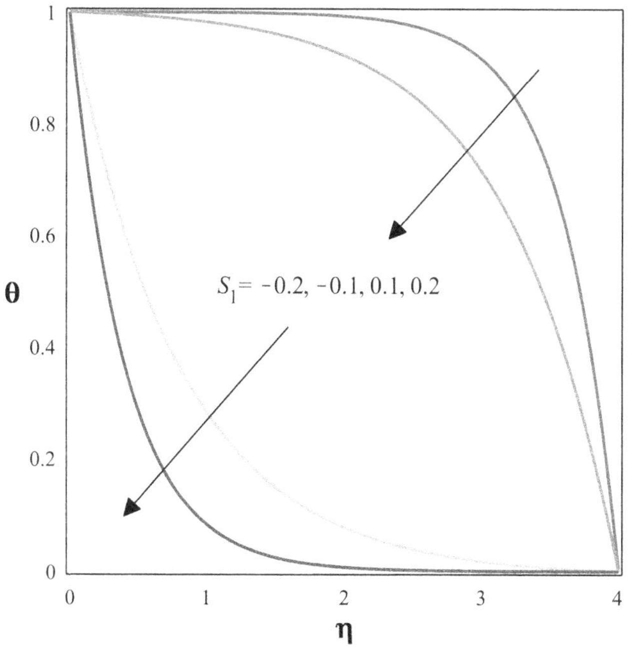

Figure 13.3 Varying $\theta(\eta)$ for several values of S_1.

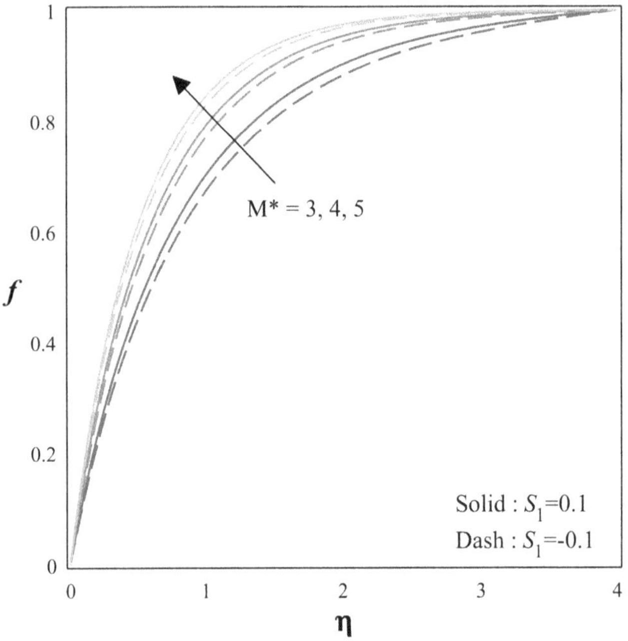

Figure 13.4 Varying $f(\eta)$ for improvement of $M*$.

This outcome shows that in the case of suction with $M*$ leads to an increase in the liquid rate than in the injection case.

Figure 13.5 demonstrates the effect of R_1 on the $\theta(\eta)$ profile in both $(S_1 > 0)$ and $(S_1 < 0)$ instances. It is noticed that when R_1 increases, the $\theta(\eta)$ improves in suction case. This result shows that enhancement the R_1 will enhance thermal dissipation and leads to more favorable thermal circulation across the diverging channel. In the case of injection, reverse nature is seen. The $\theta(\eta)$ deteriorates with escalation in R_1, this means that existence of injection drops the temperature dissipation, leads to less temperature distribution.

Figures 13.6 and 13.7 show how ϕ_3 affects $f(\eta)$ and $\theta(\eta)$ profiles in both $(S_1 > 0)$ and $(S_1 < 0)$ instances. When ϕ_3 increases, the $f(\eta)$ and $\theta(\eta)$ will changes its profile based on the chosen case.

The increase in ϕ_3 causes a drop in both the $f(\eta)$ and $\theta(\eta)$ profiles in the injection instance. This decrease is due to the presence of solid particles that obstruct the circulation and heat transmission processes. The higher the concentration of solid particles in the fluid, the more difficult it is for the fluid to flow smoothly, resulting in lower velocity. Furthermore, the presence of solid particles works as an insulator, decreasing heat transmission and resulting in lower temperatures. In the suction situation, however, an increase in ϕ_3 improves both the velocity and temperature profiles.

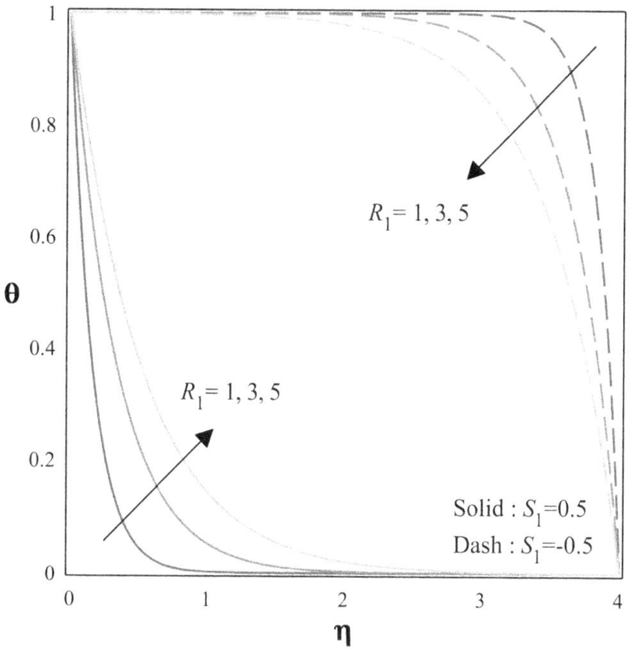

Figure 13.5 Varying $\theta(\eta)$ for several values of R_1.

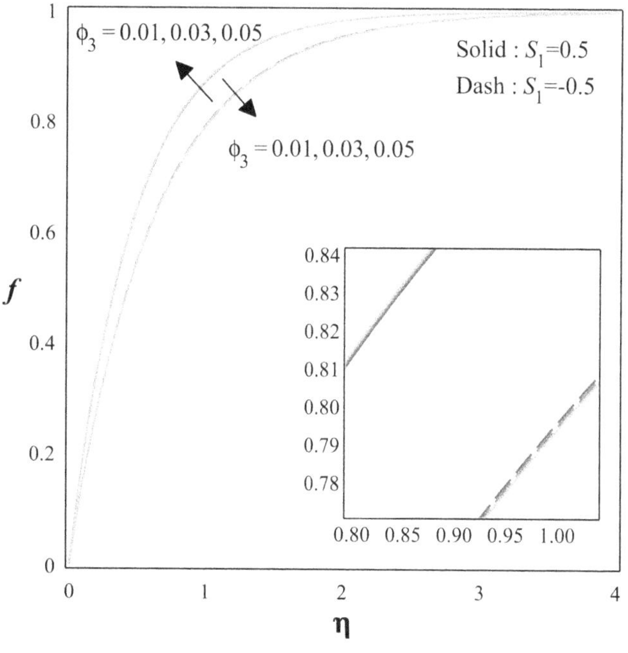

Figure 13.6 Varying $f(\eta)$ for several values of ϕ_3.

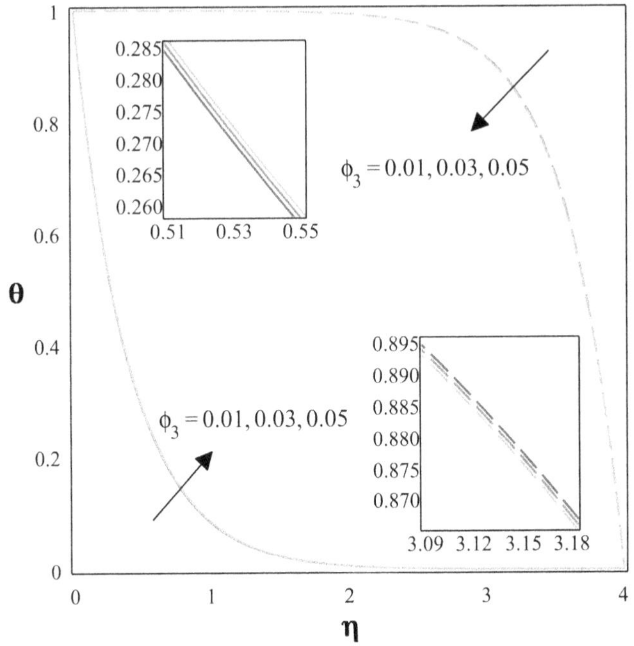

Figure 13.7 Varying $\theta(\eta)$ for several values of ϕ_3.

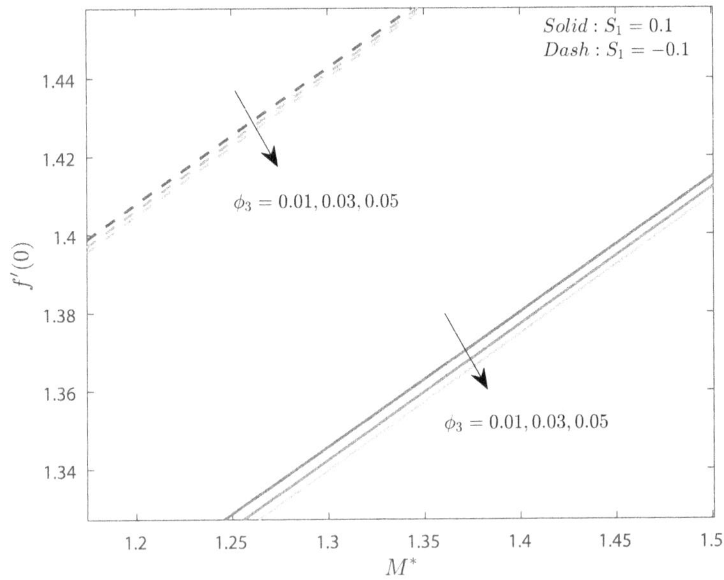

Figure 13.8 Varying Skin friction coefficient $f'(0)$ for several values of ϕ_3 with $M*$.

This behavior is caused by the suction effect, which draws fluid, including solid particles, into the channel. The larger the solid volume percentage, the denser the flow, and the higher the fluid velocities. Furthermore, the solid particles improve heat transmission by boosting conduction and convection heat transfer, which results in greater temperatures.

Figure 13.8 depicts the skin friction coefficient $f'(0)$ for various M^* and ϕ_3 values in both $(S_1 > 0)$ and $(S_1 < 0)$ instances. In both circumstances, an increase in these two constraints results in a decrease in the skin friction coefficient. A greater value of M^* indicates a stronger magnetic field, which can impact liquid flow and reduce flow resistance, resulting in a lower $f'(0)$. Similarly, raising the ϕ_3 improves flow characteristics by increasing density and facilitating smoother flow, resulting in a lower skin friction coefficient. Furthermore, the skin friction coefficient is clearly smaller in $(S_1 > 0)$ situation than in $(S_1 < 0)$ scenario. This is due to the differences in flow behavior between $(S_1 > 0)$ and $(S_1 < 0)$ cases. Suction encourages liquid flow into the channel, lowering resistance and friction, whereas injection causes opposing flow, increasing friction and hence larger skin friction coefficient.

Figure 13.9 displays the $\theta'(0)$ for various R_1 and ϕ_3 values in both $(S_1 > 0)$ and $(S_1 < 0)$ cases. When evaluating changes in these two aspects, it is discovered that the Nusselt number displays different behaviors depending on the suction or injection scenarios. In the suction situation, an increase in R_1

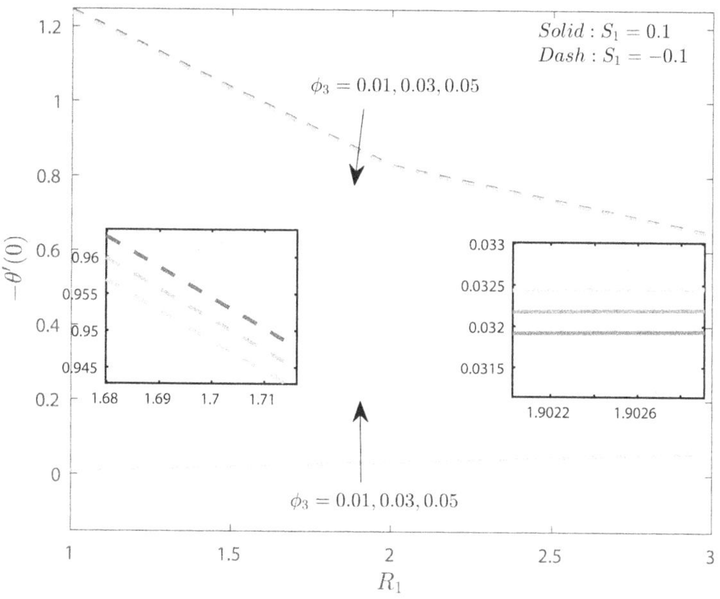

Figure 13.9 Varying Nusselt number coefficient $-\theta'(0)$ for several values of ϕ_3 with R_1.

and ϕ_3 results in a drop in $\theta'(0)$. This behavior can be due to enhanced heat transfer systems. A greater R_1 encourages more effective TR, which aids in temperature disposal and reduces temperature gradients. Furthermore, increasing ϕ_3 improves convective heat transmission by increasing the contact surface area between the liquid and solid particles. These combined impacts result in a better heat transfer rate and a lower $\theta'(0)$. In the injection instance, a rise in the $\theta'(0)$ is caused by an improvement in R_1 and ϕ_3. This is due to the opposite flow created by injection, which accelerates the convective thermal transfer process. The increased R_1 also leads to improved heat dissipation, which raises the $\theta'(0)$.

13.5 FINAL REMARKS

The current study presents a comprehensive analysis of the circulation and thermal transfer of ternary-based nanoliquid flow across a divergent channel by incorporating magnetic and TR. The investigation's results provide insightful understandings of how such systems behave, resulting in various original conclusions. The enhanced suction/injection constraint will improve the velocity while declining the temperature. The suction case shows higher velocity than the injection case. The magnetic constraint contributes to the velocity profile. In the injection instance, the inclusion of solid volume fraction lowered the velocity and temperature profiles but improved them in the suction case. The skin friction coefficient reduced when the magnetic parameter and solid volume percentage were improved, and it was typically lower in the suction scenario. When the radiation parameter and solid volume percentage are improved, the Nusselt number will be reduced in the suction case but increased in the injection case. These important discoveries have important ramifications for improving the design and functioning of divergent channel structures in diverse ternary HNF uses enabling improved functionality and effectiveness.

Divergent channel structures and ternary HNFs are important concepts in many domains, and the findings of this work have extensive potential applications in these areas. Thermal supervision, which encompasses exchangers of heat, electrically powered cooling equipment, and solar thermal power plants, is one such field. The results can help these systems be improved by increasing the rates of heat transmission and overall effectiveness. For academics, technologists, and professionals operating in these fields, this study offers useful insights that help them make defensible choices and create new approaches.

NOMENCLATURE

B	Variable magnetic field
B_0	Magnetic field $((\mathrm{kg}\Omega/\mathrm{s})^{0.5}/\mathrm{m})$

C_p	Specific heat (m²/s²/K)
Cf_x	Local skin friction
k	Thermal conductivity $\left(kgms^{-3}K^{-1}\right)$
k^*	Absorption coefficient $\left(m^{-1}\right)$
l	Length (m)
M	Magnetic field parameter
Nu_x	Nusselt number
P	Pressure gradient (kg/m/s²)
Pr	Prandtl number
Q	Constant (m²/s)
q_r	Nonlinear thermal radiation
R	Radiation parameter
Re_x	Reynolds number
$S1$	Mass suction/blowing parameter
T	Temperature (K)
T_w	The fixed temperature at the channel wall (K)
T_∞	Ambient temperature (K)
U	Variable free stream velocity (m²/s)
U^*	Characteristic velocity (m²/s)
V_w	Suction/injection velocity (ms)
u and v	Velocity components (m/s)
x and y	Coordinates (m)

GREEK SYMBOLS

α	Angle between channels
η	Similarity variable
θ	Temperature parameter
μ	Viscosity (kg/m/s)
ρ	Density (kg/m³)
σ	Electrical conductivity (per Ω/m)
σ^*	Stefan–Boltzmann coefficient (kg/s³/K⁴)
υ	Kinematic fluid viscosity (m²/s)
ϕ	Solid volume fraction

SUBSCRIPT

f	Base fluid
$thnf$	Ternary hybrid nanofluid

APPENDIX

Dimensional analysis of dimensionless constraints.

$$\text{Magnetic field parameter: } M^* = \frac{\sigma_f B_0^2 \alpha l^2}{\rho_f Q}.$$

$$\sigma_f = \text{per } \Omega/\text{m}, \quad B_0 = \sqrt{\frac{\text{kg}\Omega}{\text{s}}}/\text{m}, \quad l^2 = \text{m}^2, \quad \rho_f = \text{kg/m}^3, \quad Q = \text{m}^2/\text{s}.$$

$$M^* = \frac{\sigma_f B_0^2 \alpha l^2}{\rho_f Q} = \frac{\text{per } \Omega/\text{m} \times \text{kg}\Omega/\text{m} \times \text{m}^2}{\text{kg/m} \times \text{s} \times \text{m}^2/\text{s}} = 1.$$

$$\text{Prandtl number: } P_r = \frac{\left(C_p\right)_f \mu_f}{k_f}.$$

$$\mu_f = \text{kg/m/s}, \quad k_f = \text{kg/s} \cdot \text{m/K}, \quad \left(C_p\right)_f = \text{m}^2/\text{K/s}^2.$$

$$P_r = \frac{\mu_f \left(C_p\right)_f}{k_f} = \frac{\text{kg/m/s} \times \text{m}^2/\text{s/K}}{\text{kgm/s}^3/\text{K}} = 1.$$

$$\text{Radiation parameter: } R_1 = \frac{4\sigma^* T_\infty^3}{k^* k_f}.$$

$$\sigma^* = \text{kg/s}^3/\text{K}^4, \quad k^* = \text{per m}, \quad k_f = \text{kgm/s}^3/\text{K}, \quad T_\infty = K.$$

$$R_1 = \frac{4\sigma^* T_\infty^3}{k^* k_f} = \frac{4 \times \text{kg/s}^3/\text{K}^4 \times \text{K}^3}{\text{per m} \times \text{kgm/s}^3/\text{K}} = 4.$$

REFERENCES

[1] Choi, S. U., & Eastman, J. A. (1995). *Enhancing Thermal Conductivity of Fluids with Nanoparticles* (No. ANL/MSD/CP-84938; CONF-951135-29). Argonne National Lab. (ANL), Argonne, IL.

[2] Khan, U., Waini, I., Zaib, A., Ishak, A., & Pop, I. (2022). MHD mixed convection hybrid nanofluids flow over a permeable moving inclined flat plate in the presence of thermophoretic and radiative heat flux effects. *Mathematics*, *10*(7), 1164.

[3] Shah, N. A., Animasaun, I. L., Wakif, A., Koriko, O. K., Sivaraj, R., Adegbie, K. S., ... & Prasad, K. V. (2020). Significance of suction and dual stretching on the dynamics of various hybrid nanofluids: Comparative analysis between type I and type II models. *Physica Scripta*, *95*(9), 095205.

[4] Prasannakumara, B. C., Madhukesh, J. K., & Ramesh, G. K. (2023). Bioconvective nanofluid flow over an exponential stretched sheet with thermophoretic particle deposition. *Propulsion and Power Research*, 12, 284–296.

[5] Prabakaran, R., Eswaramoorthi, S., Loganathan, K., & Sarris, I. E. (2022). Investigation on thermally radiative mixed convective flow of carbon nanotubes/Al_2O_3 nanofluid in water past a stretching plate with joule heating and viscous dissipation. *Micromachines*, *13*(9), 1424.

[6] Madhukesh, J. K., Ramesh, G. K., Shehzad, S. A., Chapi, S., & Prabhu Kushalappa, I. (2023). Thermal transport of MHD Casson-Maxwell nanofluid between two porous disks with Cattaneo-Christov theory. *Numerical Heat Transfer, Part A: Applications*, 1–16. DOI: 10.1080/10407782.2023.2214322.

[7] Ramesh, G. K., Shehzad, S. A., & Izadi, M. (2020). Thermal transport of hybrid liquid over thin needle with heat sink/source and Darcy-Forchheimer porous medium aspects. *Arabian Journal for Science and Engineering*, *45*, 9569–9578.

[8] Madhukesh, J. K., Ramesh, G. K., Shehzad, S. A., Rauf, A., & Omar, M. (2023). A microstructural slip analysis of radiative thermophoretic flow of ternary nanofluid flowing through porous medium. *Physica Scripta*, *98*(6), 065213.

[9] Patel, S. K., Lavadiya, S. P., Parmar, J., Ahmed, K., Taya, S. A., & Das, S. (2022). Design and fabrication of reconfigurable, broadband and high gain complementary split-ring resonator microstrip-based radiating structure for 5G and WiMAX applications. *Waves in Random and Complex Media*, 1–31.

[10] Shukla, S., Sharma, R. P., Punith Gowda, R. J., &Prasannakumara, B. C. (2023). Elastic deformation effect on carboxymethyl cellulose water-based (TiO_2-Ti_6Al_4V) hybrid nanoliquid over a stretching sheet with an induced magnetic field. *Numerical Heat Transfer, Part A: Applications*, *84*, 1–15.

[11] Animasaun, I. L., Al-Mdallal, Q. M., Khan, U., &Alshomrani, A. S. (2022). Unsteady water-based ternary hybrid nanofluids on wedges by bioconvection and wall stretching velocity: Thermal analysis and scrutinization of small and larger magnitudes of the thermal conductivity of nanoparticles. *Mathematics*, *10*(22), 4309.

[12] Manohar, G. R., Venkatesh, P., Gireesha, B. J., Madhukesh, J. K., & Ramesh, G. K. (2022). Performance of water, ethylene glycol, engine oil conveying SWCNT-MWCNT nanoparticles over a cylindrical fin subject to magnetic field and heat generation. *International Journal of Modelling and Simulation*, *42*(6), 936–945.

[13] Ferdows, M., Ramesh, G. K., & Madhukesh, J. K. (2023). Magnetohydrodynamic flow and Hall current effects on a boundary layer flow and heat transfer over a three-dimensional stretching surface. *International Journal of Ambient Energy*, *44*(1), 938–946.

[14] Swain, K., Animasaun, I. L., & Ibrahim, S. M. (2022). Influence of exponential space-based heat source and Joule heating on nanofluid flow over an elongating/shrinking sheet with an inclined magnetic field. *International Journal of Ambient Energy*, *43*(1), 4045–4057.

[15] Khan, U., Zaib, A., Ishak, A., Waini, I., Raizah, Z., & Galal, A. M. (2022). Analytical approach for a heat transfer process through nanofluid over an irregular porous radially moving sheet by employing KKL correlation with magnetic and radiation effects: Applications to thermal system. *Micromachines*, *13*(7), 1109.
[16] Madhukesh, J. K., Alam, M. M., Varun Kumar, R. S., Arasaiah, A., Ahmad, I., Gorji, M. R., & Prasannakumara, B. C. (2022). Exploring magnetic dipole impact on Maxwell hybrid nanofluid flow over a stretching sheet. *Proceedings of the Institution of Mechanical Engineers, Part E: Journal of Process Mechanical Engineering.* doi:10.1177/09544089211073267.
[17] Das, S., Ali, A., Jana, R. N., & Makinde, O. D. (2022). EDL impact on mixed magneto-convection in a vertical channel using ternary hybrid nanofluid. *Chemical Engineering Journal Advances*, *12*, 100412.
[18] Ramesh, G. K., Madhukesh, J. K., Shah, N. A., & Yook, S. J. (2023). Flow of hybrid CNTs past a rotating sphere subjected to thermal radiation and thermophoretic particle deposition. *Alexandria Engineering Journal*, *64*, 969–979.
[19] Abbas, A., Ashraf, M., Sarris, I. E., Ghachem, K., Labidi, T., Kolsi, L., & Ahmad, H. (2023). Numerical simulation of the effects of reduced gravity, radiation and magnetic field on heat transfer past a solid sphere using finite difference method. *Symmetry*, *15*(3), 772.
[20] Zeeshan, Khan, I., Eldin, S. M., Islam, S., & Khan, M. U. (2023). Two-dimensional nanofluid flow impinging on a porous stretching sheet with nonlinear thermal radiation and slip effect at the boundary enclosing energy perspective. *Scientific Reports*, 13(1), 5459.
[21] Khan, U., Zaib, A., Ishak, A., Waini, I., & Pop, I. (2022). Mixed convection flow of water conveying graphene oxide nanoparticles over a vertical plate experiencing the impacts of thermal radiation. *Mathematics*, *10*(16), 2833.
[22] Alzahrani, H. A., Alsaiari, A., Madhukesh, J. K., Naveen Kumar, R., & Prasanna, B. M. (2022). Effect of thermal radiation on heat transfer in plane wall jet flow of Casson nanofluid with suction subject to a slip boundary condition. *Waves in Random and Complex Media*, 1–18.
[23] Alharbi, K. A. M., Farooq, U., Waqas, H., Imran, M., Noreen, S., Akgül, A., ... & Abbas, K. (2023). Numerical solution of Maxwell-Sutterby nanofluid flow inside a stretching sheet with thermal radiation, exponential heat source/sink, and bioconvection. *International Journal of Thermofluids*, *18*, 100339.
[24] Ramesh, G. K., Madhukesh, J. K., Shehzad, S. A., & Rauf, A. (2022). Ternary nanofluid with heat source/sink and porous medium effects in stretchable convergent/divergent channel. *Proceedings of the Institution of Mechanical Engineers, Part E: Journal of Process Mechanical Engineering.* doi:10.1177/09544089221081344.
[25] Xu, Y. J., Nazeer, M., Hussain, F., Khan, M. I., Hameed, M. K., Shah, N. A., & Chung, J. D. (2021). Electro-osmotic flow of biological fluid in divergent channel: Drug therapy in compressed capillaries. *Scientific Reports*, *11*(1), 23652.
[26] Banerjee, A., Mahato, S. K., Bhattacharyya, K., &Chamkha, A. J. (2021). Divergent channel flow of Casson fluid and heat transfer with suction/blowing and viscous dissipation: Existence of boundary layer. *Partial Differential Equations in Applied Mathematics*, 4, 100172.

[27] Nazeer, M., Hussain, A., & Hameed, M. K. (2022). Impact of nano metallic particles and magnetic force on multi-phase flow of third-grade fluid in divergent channel: Analytical study. *International Journal of Modelling and Simulation*, 1–12.
[28] Verma, A. K., Gautam, A. K., Bhattacharyya, K., & Sharma, R. P. (2021). Existence of boundary layer nanofluid flow through a divergent channel in porous medium with mass suction/injection. *Sādhanā*, *46*(2), 98.
[29] Isaev, S. A., Baranov, P. A., Guvernyuk, S. V., & Zubin, M. A. (2002). Numerical and physical modeling of turbulent flow in a divergent channel with a vortex cell. *Journal of Engineering Physics and Thermophysics*, *75*(2), 269–276.
[30] Anantha Kumar, K., Sandeep, N., Sugunamma, V., & Animasaun, I. L. (2020). Effect of irregular heat source/sink on the radiative thin film flow of MHD hybrid ferrofluid. *Journal of Thermal Analysis and Calorimetry*, *139*, 2145–2153.
[31] Ali, K., Reddy, Y. R., & Shekar, B. C. (2023). Thermo-fluidic transport process in magnetohydrodynamic couette channel containing hybrid nanofluid. *Partial Differential Equations in Applied Mathematics*, 7, 100468.
[32] Layek, G.C., Kryzhevich, S.G., Gupta, A.S. *& Reza* M. (2013) Steady magnetohydrodynamic flow in a diverging channel with suction or blowing. *Zeitschrift für Angewandte Mathematik und Physik 64*, 123–143. https://doi.org/10.1007/s00033-012-0225-9

Index

For Product Safety Concerns and Information please contact our EU representative GPSR@taylorandfrancis.com
Taylor & Francis Verlag GmbH, Kaufingerstraße 24, 80331 München, Germany

www.ingramcontent.com/pod-product-compliance
Lightning Source LLC
Chambersburg PA
CBHW060407220326
41598CB00023B/3054
9781032712062